# Lecture Notes in Electrical E

## Volume 753

The book series *Lecture Notes in Electrical Engineering* (LNEE) publishes the latest developments in Electrical Engineering - quickly, informally and in high quality. While original research reported in proceedings and monographs has traditionally formed the core of LNEE, we also encourage authors to submit books devoted to supporting student education and professional training in the various fields and applications areas of electrical engineering. The series cover classical and emerging topics concerning:

- Communication Engineering, Information Theory and Networks
- Electronics Engineering and Microelectronics
- Signal, Image and Speech Processing
- Wireless and Mobile Communication
- Circuits and Systems
- Energy Systems, Power Electronics and Electrical Machines
- Electro-optical Engineering
- Instrumentation Engineering
- Avionics Engineering
- Control Systems
- Internet-of-Things and Cybersecurity
- Biomedical Devices, MEMS and NEMS

For general information about this book series, comments or suggestions, please contact leontina.dicecco@springer.com.

To submit a proposal or request further information, please contact the Publishing Editor in your country:

**China**

Jasmine Dou, Editor (jasmine.dou@springer.com)

**India, Japan, Rest of Asia**

Swati Meherishi, Editorial Director (Swati.Meherishi@springer.com)

**Southeast Asia, Australia, New Zealand**

Ramesh Nath Premnath, Editor (ramesh.premnath@springernature.com)

**USA, Canada:**

Michael Luby, Senior Editor (michael.luby@springer.com)

**All other Countries:**

Leontina Di Cecco, Senior Editor (leontina.dicecco@springer.com)

**\*\* This series is indexed by EI Compendex and Scopus databases. \*\***

More information about this series at http://www.springer.com/series/7818

Girolamo Di Francia · Corrado Di Natale
Editors

# Sensors and Microsystems

## Proceedings of the AISEM 2020 Regional Workshop

Co-edited by B. Alfano, S. De Vito, A. Del Giudice,
E. Esposito, S. Fattoruso, S. Ferlito, F. Formisano,
E. Massera, M. L. Miglietta, T. Polichetti

 Springer

*Editors*
Girolamo Di Francia
ENEA
Portici, Italy

Corrado Di Natale
Department of Electronic Engineering
University of Rome Tor Vergata
Rome, Italy

ISSN 1876-1100        ISSN 1876-1119  (electronic)
Lecture Notes in Electrical Engineering
ISBN 978-3-030-69553-8        ISBN 978-3-030-69551-4  (eBook)
https://doi.org/10.1007/978-3-030-69551-4

This Springer imprint is published by the registered company Springer Nature Switzerland AG
The registered company address is: Gewerbestrasse 11, 6330 Cham, Switzerland

# Preface

This book contains the papers presented, in February 2020, at the AISEM (Italian National Association on Sensors and Microsystems) regional workshop, held in the Portici ENEA Research Centre, Italy, and organized by ENEA, the Italian National Agency for New Technologies, Energy and Sustainable Economic Development.

Perhaps more than for the quality of the papers presented and collected in this book, the workshop will be recalled for its simultaneity with the COVID-19 pandemic that furiously attacked all the world and that, at the end of 2020, we all are still fighting. Up to now, more than 1.5 million people have already died in consequence, and if humans will be able to win the virus it will only be either for the valiant work of the healthcare personnel and for the incredible work that researchers all over the world have done, in less than one year, to get a vaccine. One time more this proves how relevant scientific research is and what important for everybody's life can our work be.

The 34 papers here presented are organized, as it is a tradition for AISEM meetings, in materials and processing technologies, sensor and microsystem devices, sensor system and device applications. It was not easy to finalize this book: the coronavirus attacks the body but also our spirit and as the Chairman I feel that I have to thank both the authors for the quality of the works presented and the reviewers, for their attention and prompt availability. But, finally, let me really and sincerely deeply thank also all ENEA co-editors: without their work, probably this book would not have seen the light.

This year, for the worldwide sensor community, will be also recalled for the death of Prof. Arnaldo D'Amico, pioneer of sensors science, mentor of many of us and AISEM founder. A tribute, written by Corrado Di Natale, presently AISEM Chair, is also included in this volume. I met Arnaldo while living, with him, the first years of the AISEM Conference. AISEM, his creation, the one that I lived from its first steps, reveals by itself, a lot of what its creator was. Arnaldo had wanted that in this association and in its annual conference, different knowledge, even very far one from another, could gather in a cultural approach that is not common for scientific meetings. In this respect, perhaps even without being conscious of that, Arnaldo was a revolutionary because he wanted to look where others did not want or could look. There was in him and therefore there is in AISEM, a tension towards innovation, to look beyond, to mix things with the awareness that from this apparent confusion the positive progress of the times can arise, the look ahead that only humans and never machines, can show. This is what I think Arnaldo leaves us: the certainty that in the ambition to know, even in the awareness of never being able to reach the absolute, we work for the best, for the goodness of everyone.

Portici, Italy                                                                Girolamo Di Francia
                                                                                    Workshop Chair

# A Tribute to Prof. Arnaldo D'Amico

Prof. Arnaldo D'Amico was born in Bologna, 16 March 1940, and died in Rome, 11 November 2020. He got doctor degrees (laurea) in physics and then in electronic engineering from the University of Rome La Sapienza. He began his scientific career as a researcher at the Institute of Solid State Electronics of the National Research Council (CNR) until he became full professor of electronics at the University of L'Aquila and, since 1990, at the University of Rome Tor Vergata. He has been a pioneer of sensors science: he was a reference for the international community and, in particular, in Italy.

In Italy, he promoted the research in sensors directing the first national project dedicated to sensors and from 1996 founded and directed for several years the Associazione Italiana Sensori e Microsistemi (AISEM). In Europe, he has been among the founders of the series of conferences Eurosensors, of which he directed for years the steering committee, and, since the first issue, he has been a member of the editorial board of the journals *Sensors and Actuators A and B.*

He has been a passionate teacher. At the University of Rome Tor Vergata besides teaching electronic devices, he designed courses on sensors for students of electronic and medical engineering.

His research interests were concerned with any aspect of sensors science, from the design of sensors to the development of sensor systems. At the University of Rome Tor Vergata, he founded and directed, until the retirement, an interdisciplinary research group. He published more than 500 papers, and among his scientific achievements it is worth to mention his pioneering work in the application of gas sensors to the analysis of volatile metabolites.

He has been a supervisor of dozens of Ph.D. and master thesis. Characterized by distinct humanity and empathy, he has been a mentor of many researchers and professors in Italy and abroad.

He leaves a great legacy and in those that had the chance to work with him is a great feeling of gratitude for the gift of his person, for what he taught and for his kindness.

Corrado Di Natale

# Contents

# Smart City Platform: Scalability, Interoperability and Replicability Platform to Manage Urban Applications

M. Chinnici, G. Ponti, and G. Santomauro

**Abstract** In a smart city environment, the explosive growth in the volume, speed, and variety of data being produced every day requires a continuous increase in the processing speeds of servers and entire network infrastructures, platforms as well as new resource management models. This poses significant challenges for data-intensive and high-performance computing, i.e., how to turn enormous datasets into valuable information and meaningful knowledge efficiently. In this work, the authors propose an approach and describe a methodology and a modular and scalable multi-layered ICT platform called ENEA Smart City Platform (ENEA-SCP) to address the problem of cross-domain interoperability in the context of smart city applications and for offering services to the users (e.g. public administration, citizens, providers).

**Keywords** Smart city · Big data · ICT platform · Interoperability · IoT

## 1 Introduction

The variety of sources complicates the task of context data management such as data derives from, resulting in different data formats, with varying storage, transformation, delivery, and archiving requirements. At the same time, rapid responses are needed for real-time applications. With the emergence of cloud infrastructures and platforms, achieving highly scalable data management in such contexts is a critical problem, as the overall urban application performance is highly dependent on the properties of the data management service (Fig. 1). Hence, continuously developing and adopting

M. Chinnici
ENEA, CR Casaccia, Via Anguillarese 301, 00123 Roma, Italy
e-mail: marta.chinnici@enea.it

G. Ponti (✉) · G. Santomauro
ENEA, CR Portici, Piazzale Enrico Fermi 1, 80055 Napoli, Italy
e-mail: giovanni.ponti@enea.it

G. Santomauro
e-mail: giuseppe.santomauro@enea.it

G. Di Francia and C. Di Natale (eds.), *Sensors and Microsystems*,
Lecture Notes in Electrical Engineering 753,
https://doi.org/10.1007/978-3-030-69551-4_1

1

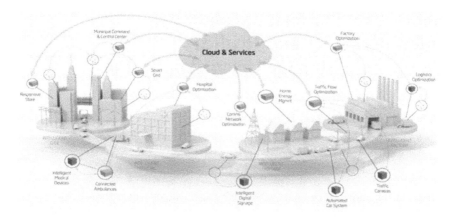

**Fig. 1** An example of an integrated urban district [2]

ICT technologies to create and use platforms for government, business and citizens can communicate and work together and provide the necessary connections between the networks that are the base for the services of the smart city [1].

The main features of a generic Smart City Platform (SCP) are in the following [3]

- Make data, information, people and organizations smarter;
- Redesign the relationships between government, private sector, non-profits, communities and citizens;
- Ensure synergies and interoperability within and across city policy domains and systems (e.g. transportation, energy, education, health and care, utilities, etc.);
- Drive innovation, for example, through so-called open data, living labs and tech-hub.

The ENEA-SCP is implemented following the Software as a Service (SaaS) paradigm, exploiting cloud computing facilities to ensure flexibility and scalability. Interoperability and communication are addressed employing web services, and data format exchange is based on the JSON data format. By taking into account these guidelines as references, this work provides a description of the SCP developed by ENEA [4] and its potential use for smart and IoT city applications. In summary, the ENA-SCP aim is to interact with most of components at the urban district level by a flexible and multipurpose data format.

The paper is organized as follows. Section 1 provides an introduction to the addressed problem; in Sect. 2, a brief outline of the reference SCP model in terms of Interoperability, Scalability and Replicability issues for SCP is described; finally, Sect. 3 concludes the paper and gives some ideas for future works.

## 2 ENEA Smart City Platform

The solution provided by ENEA-SCP to exploit potentials in Smart City environments is based on four fundamental concepts:

- Open Data
- Interoperability
- Scalability
- Replicability.

These four aspects, whose related issues were tacked in ENEA-SCP development, provide a reference framework of modular specifications [3] for stakeholders willing to implement ICT platforms in order to exploit the Smart City vision potentials and therefore offer new services for the citizen.

### 2.1 Open Data: UrbanDataset

The starting point for providing interoperability into the SCP development has been to define an open data format able to guarantee a flexible and, at the same time, accurate communication between SCP and all data producers/consumers (named *Solutions*). For such a crucial issue, a JSON based data format, named *UrbanDataset* (*UD*), has been defined.

A more structured and detailed classification of the concepts associated with properties and UDs has been outlined, to facilitate their reuse and search within an ontology. This classification, starting from standard taxonomies (such as ISO 37120 [5]) and the requirements of the UrbanDatasets themselves, clarifies the meaning of the property, explaining the type of represented concept, whether they are general meta-information (identifiers, descriptions, formats, etc.) or quantities related to specific Smart City application domains (telecommunications, energy, urban planning, etc.).

### 2.2 Interoperability

One of the most innovative aspects of the proposed approach is related to the information and the semantic interoperability levels. The standard issue, referring to interoperability by use of shared data formats, is how to find the correct balance between too prescriptive specifications (which guarantee interoperability, but the risk to inhibit innovation) and more elastic specifications (which have a lot of potential deficit concerning real interoperability). This problem becomes urgent in a context, like Smart City, with a lot of interacting heterogeneous systems. The proposed approach is focused on overcoming this issue; indeed, it has a very light and elastic format at

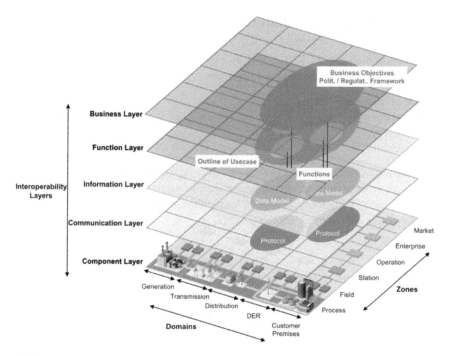

**Fig. 2** Smart grid operation layers structured for universal interoperability [6]

information level able to represent an extensive set of data, moving at the semantic level the strict definition of the data. The underlying idea is that this light approach can be easily applied also on existing systems with just small intervention on them.

In detail, five sub-specifications concerning the interoperability levels have been implemented and validated. The list below shows the implementation progress made for each of the interoperability levels (Fig. 2) with the related specifications:

1. **Functional**: design of the SCP Reference Architecture and functionalities;
2. **Collaboration**: definition of the Registry and the basic GUI for the management of the collaboration;
3. **Semantic**: definition of the Ontology for centralized interpretation;
4. **Information**: data model managed by the SCP and XML and JSON formats to represent the data;
5. **Communication**: definition of the patterns and interfaces supported for data transport.

## 2.3 Scalability

The ENEA Smart City Platform exploits computational resources of the ENEA-GRID infrastructure [4], as it is deployed in the cloud-hosted in the Portici Research

Center site. The creation of a customized environment ENEA-cloud-based platform is possible thanks to the virtualization technologies of VMWARE platform, which allows hosting the management, the transportation and the processing of project data services, ensuring their availability and protection over time. More in detail, the SCP is composed of six Virtual Machines (VMs), and each of them hosts a component with a specific role (Fig. 3). There are two access points, one for the Machine to Machine (M2M) interaction and one for the Human–Machine Interface (HMI) interaction, respectively named *scp-ws* (as web-services provider) and *scp-gui* (as web-accessible GUI). The *scp-id* is used as an identity provider while the *scp-registry* is the orchestrator that manages the production and the access of data by means a relational database. The *scp-ud* is the VMs that stores data by exploiting a No-SQL database. Finally, there is a private VM, the *scp-ws-inner*, which is similar to *scp-ws* but is used only for HMI transactions.

This architecture presents the peculiarity of both horizontal and vertical scalability; as the number of VMs and/or processing resources can be varied according to the volume and to the throughput of the handled data. Also, the flexibility guaranteed by this cloud solution, combined with integration in ENEAGRID resources, provides an immensely powerful and functional Smart City Platform, which exploits the main features of the whole environment.

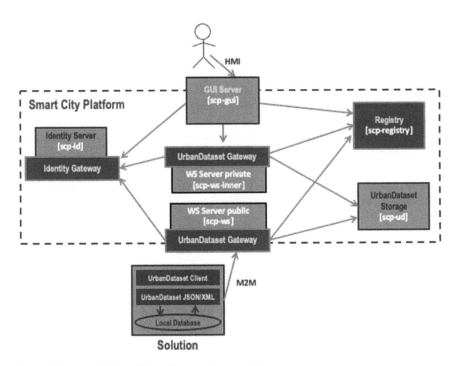

**Fig. 3** Scheme of ENEA-SCP architecture based on VMs

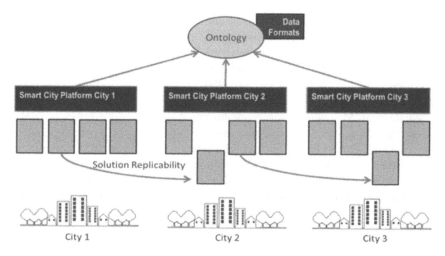

**Fig. 4** Example of the replicability solution

## 2.4  *Replicability*

The open-source solutions have only been adopted to ensure the SCP replicability'
system and the components have been developed in such a way; then they can be
easily reinstalled in a new pool of virtual machines.

In particular, the execution software of each VM is parametrized by means of
a configuration file that allows defining proper settings for a specific installation
(e.g. the name of the virtual machines or particular access credentials to the various
systems). This technological solution allows replicating overall SCP implementation
in several and different urban districts, both server-side (SCP replicability) and client-
side (solution replicability) as depicted, e.g., in Fig. 4.

The replicability characteristic of the SCP allows its versatile use both in terms
of management of the vast amount of the smart city applications and for offering
services to the potential users such as public administration, citizens, providers.

## 3  Conclusions

In this work, the authors presented the ICT ENEA Smart City Platform (SCP) able
to interact with different stakeholders at the urban district level by a flexible and
multipurpose data format. The ENEA-SCP milestones are the opportunity to scale the
computational resources according to requests; interoperate, using five specification
levels, with all the interesting parts; and replicate all the components in the different
city context. For the future works, the aims of the authors are: to develop a "light"
ENEA-SCP where all components described in Sect. 2 (as in Fig. 3) are gathered in a

single Virtual Machine (in this way the deployment of a new SCP became easier); to create a unique identity component (*scp-id*) for all SCP instances; finally, to create a "National SCP" (*interSCP* or *iSCP*) with the goal of collecting and elaborating data from all SCPs running at the urban level.

# References

1. Chinnici M, De Vito S (2018) IoT meets opportunities and challenges: edge computing in deep urban environment. In: Dependable IoT for human and industry. Modeling, architecting, implementation. River Publishers series in information science and technology
2. Novelli C, Frascella A et al (2017) Piattaforma ICT per la gestione dello smart district. Report RdS/PAR2016/001
3. Brutti A, De Sabbata P et al (2018) Smart city platform specification: a modular approach to achieve interoperability in smart cities. In: The internet of things for smart urban ecosystems. Springer, Berlin, pp 25–50
4. Ponti G et al (2014) The role of medium size facilities in the HPC ecosystem: the case of the new CRESCO4 cluster integrated in the ENEAGRID infrastructure. In: Proceedings of the 2014 international conference on high performance computing and simulation, HPCS 2014, art. no. 6903807, pp 1030–1033
5. ISO 37120:2018 Sustainable cities and communities—indicators for city services and quality of life
6. Smart grid reference architecture (2012). CEN-CENELEC-ETSI Smart Grid Coordination Group

# Real-Time Obstacle Detection and Field Mapping System Using LIDAR-ToF Sensors for Small UAS

**Gennaro Ariante**ⓘ**, Umberto Papa**ⓘ**, Salvatore Ponte**ⓘ**, and Giuseppe Del Core**ⓘ

**Abstract** UASs (Unmanned Aircraft Systems) are becoming increasingly popular, for both military and civil applications. They are widely used in various tasks, such as search and rescue, disaster assessment, urban traffic monitoring, 3D mapping, etc., that would be risky or impossible to perform for a human. DAA (Detect and Avoid) is a new UAS technology necessary to safely avoid obstacles or other UASs and aircrafts. In this work low-cost sensors, namely, a DAA architecture based on a LIDAR (Light Detection and Ranging), and a ToF (Time of Flight) sensor, will be installed on a small unmanned rotorcraft to estimate its distance from an obstacle and for field mapping. To correct the data from systematic errors (bias) and measurement noise, Kalman filtering and a criterion of optimal estimation have been implemented. Collected data are sent to a microcontroller (Arduino Mega 2560), which allows for low-cost hardware implementations of multiple sensors for use in aerospace applications.

**Keywords** UAS · UAV · DAA · Lidar · Sensor fusion · ToF · Kalman filter

G. Ariante (✉) · U. Papa · G. Del Core
Department of Science and Technology, University of Naples "Parthenope", Napoli, Italy
e-mail: gennaro.ariante@studenti.uniparthenope.it

U. Papa
e-mail: umberto.papa@uniparthenope.it

G. Del Core
e-mail: giuseppe.delcore@uniparthenope.it

S. Ponte
Department of Engineering, University of Studies of Campania "Luigi Vanvitelli", Aversa, CE, Italy
e-mail: salvatore.ponte@unicampania.it

© The Author(s), under exclusive license to Springer Nature Switzerland AG 2021
G. Di Francia and C. Di Natale (eds.), *Sensors and Microsystems*,
Lecture Notes in Electrical Engineering 753,
https://doi.org/10.1007/978-3-030-69551-4_2

# 1  Introduction

Research in UAS field is getting more and more attention and importance due to their wide application, both military and civilian [1]. UASs can perform missions that pose high risks to human operators, such as search and rescue, reconnaissance and strike, surveillance and monitoring in danger-prone or inaccessible sites [2].

Knowledge of the environment, mapping and localization is an important task for the remote pilots and/or autonomous flight, particularly during the different flight phases in unknown areas. In this paper we describe the preliminary steps towards the development of a DAA system with low-cost sensors, in particular LIDAR-ToF. The platform will be installed on a UAV to send information on position, detect obstacles and for field mapping [3–6]. Both sensors have typical "low-cost" characteristics, i.e. miniaturization, fast response time and good sensing range for obstacle detection. They are managed through I2C serial connection by a microcontroller Arduino Mega 2560 [7].

# 2  Theoretical Framework

## 2.1  Sensor Fusion

Multisensors data fusion is an essential task for and improved estimation of system states and parameters [2, 8]. The data fusion module implemented in this work aims to reduce uncertainties on the distance measurements from a fixed or movable object that could become an obstacle during the flight. Additional distance information is obtained thanks to LIDAR (Lidar lite v3) and ToF (VL53L0X) sensors, added to the standard instrumentation (Fig. 1), in order to perform enhanced automatic obstacle detection and distance-from-obstacle estimation around the flight area.

The LIDAR lite v3 measures distance by calculating the time delay between the transmission of a Near-Infrared laser signal and its reception after reflection from a target. This translates into distance (meters or feet) using the speed of light [9]. The VL53L0X is a new generation ToF laser-ranging module housed in the smallest package on the market today, providing accurate distance measurements whatever the target reflectance, unlike conventional technologies. It can measure absolute distances up to 2 m, setting a new benchmark in ranging performance levels, opening up various and interesting new applications [10].

## 2.2  Kalman Filtering and Gelb's Method

The algorithms used for estimation and removal of systematic errors and noise are Kalman filtering and Gelb's method [11] for sensor data fusion. The Kalman filter

Arduino Mega 2560

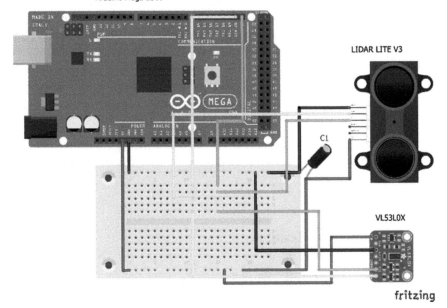

LIDAR LITE V3

VL53L0X

fritzing

**Fig. 1** Electrical scheme of the system

(KF) is a widely used quadratic state estimator for discrete linear dynamic systems perturbed by white noise ($w$), which uses measurements linearly related to the state and corrupted by white Gaussian noise [11, 12]:

$$x_k = A x_{k-1} + B u_{k-1} + w_{k-1} \tag{1}$$

$$z_k = H x_k + v_k \tag{2}$$

where $x_k$ is the state vector evaluated at time $t_k$, $A$ and $B$ are the state and input matrices, $w_k$ is the process noise vector, with covariance matrix $Q$, $z_k$ is the $k$th measurement vector, $H$ is the observation matrix, and $v_k$ is the measurement noise, with covariance matrix $R$. Starting from an initial state estimate $\hat{x}_0$ and state error covariance matrix $P_0$, the KF is based on a prediction-correction strategy, projecting forward the current state $\hat{x}_k^-$ [*a-priori* estimate, Eq. (1)] and predicting the *a-posteriori* state estimate $\hat{x}_k$ based on the current measurement weighted by a gain matrix $K_k$: (Kalman gain):

$$\hat{x}_k = \hat{x}_k^- + K_k \left( z_k - H \hat{x}_k^- \right) \tag{3}$$

$$K_k = \frac{P_k^- H^T}{H P_k^- H^T + R}; \ P_k = (I - K_k H) P_k^- \tag{4}$$

The estimation errors are provided by the element of the matrix $\boldsymbol{P}$.

Gelb's method is a simple data fusion algorithm that processes measurements to deduce a linear estimate $\hat{x}$ of the unknown quantity (i.e. distance) which, assuming random, independent and unbiased measurement errors, minimizes the mean square value of the estimation error [11]:

$$\hat{x} = \left(\frac{\sigma_{LDR}^2}{\sigma_{ToF}^2 + \sigma_{LDR}^2}\right) z_{ToF} + \left(\frac{\sigma_{ToF}^2}{\sigma_{ToF}^2 + \sigma_{LDR}^2}\right) z_{LDR} \tag{5}$$

where the variances of the measurements of the LIDAR, $z_{LDR}$, and ToF, $z_{ToF}$, are $\sigma_{LDR}^2$ and $\sigma_{ToF}^2$ respectively. It can be shown that the minimum mean square estimation error is $\left(1/\sigma_{LDR}^2 + 1/\sigma_{ToF}^2\right)^{-1}$.

## 3 Simulations and Results

To assess the operative characteristic of the sensors, simulations are performed in the obstacle detection range, chosen to be 30–180 cm. The obstacle is moved in 5-cm steps during the data acquisition sessions. Measurements were acquired for of 120 s at 4 Hz (one measurement every 250 ms, 480 samples per acquisition) (see Fig. 2).

In post-processing, mean and variance have been evaluated in each data collection for raw and filtered data (see Fig. 3).

A comparison between mean and variance of the measurements (Table 1) shows a bias in the LIDAR acquisitions, which was corrected by applying Kalman filtering

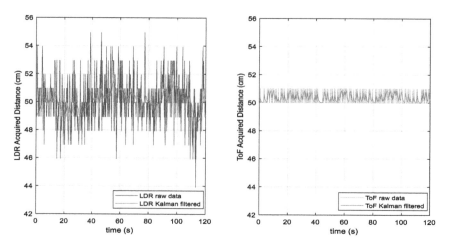

**Fig. 2** Data collection of LIDAR (left) and ToF (right) during a static test, with the obstacle at 50 cm

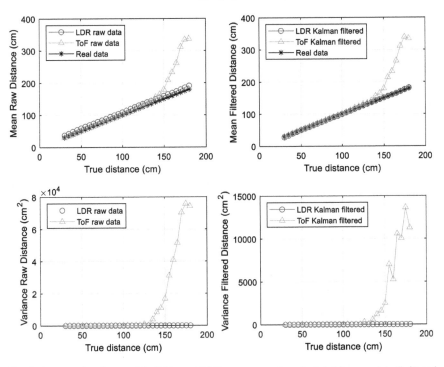

**Fig. 3** Mean and variance of the Lidar and ToF measurements, calculated for raw data (left) and filtered data (right)

**Table 1** Comparison mean and variance of some raw (R) and Kalman (K) filtered distances

| | LIDAR | | | ToF | | |
|---|---|---|---|---|---|---|
| Distance (cm) | 30 | 90 | 150 | 30 | 90 | 150 |
| – | R/K | R/K | R/K | R/K | R/K | R/K |
| Mean (cm) | 37/28 | 99/90 | 159/150 | 30/30 | 90/90 | 179/163 |
| VAR (cm$^2$) | 4.6/0.7 | 1.9/0.2 | 2.01/0.2 | 0/0 | 0.2/0 | $1.6 \times 10^4/2 \times 10^3$ |

to the raw data. ToF measurements shows good accuracy at short distances (less than 120 cm) and high errors for longer distances.

As an alternative, after estimating the sensor variances $\sigma^2_{LDR}$ and $\sigma^2_{ToF}$ from static measurements, Eq. (5) is applied to obtain the optimal estimated distance. Results are shown in Fig. 4, and in Fig. 5 and Table 2 the methodology is compared to the Kalman filtering approach.

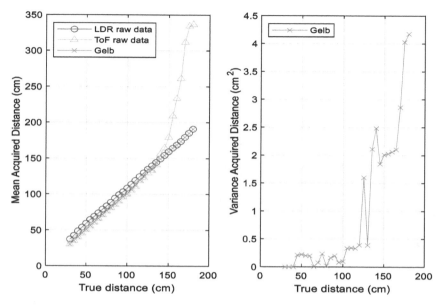

**Fig. 4** Mean and variance of the values estimated by Gelb's method

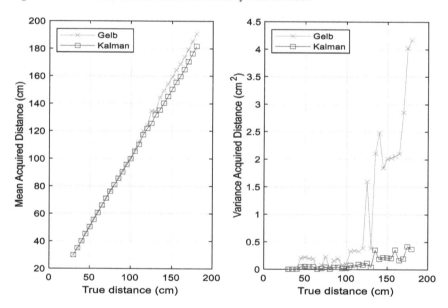

**Fig. 5** Comparison between Gelb's method with Kalman filter

| Table 2 Comparison between raw data, Gelb's data fusion and Kalman-filtered data | Raw data LDR/ToF | Gelb | Kalman |
|---|---|---|---|
| RMS (cm) | 9.1/53.6 | 5.9 | 0.8 |
| Max error (cm) | 10.8/159.4 | 10.9 | 2.2 |

## 4 Conclusion

This paper has quickly described the preliminary steps towards the implementation of a Detect-And-Avoid subsystem onboard an UAV, exploiting low-cost, commercial off-the-shelf distance measuring sensors (LIDAR and ToF), handled by easily programmable microcontrollers (Arduino Mega 2560). Laboratory simulations and experimental results on a prototype multisensor board developed by the authors show that simple data fusion techniques (linear estimation and Kalman filtering) provide improved observability, reducing the error region, broadening the baseline of the observable (distance from an obstacle in the range 30–180 cm) and helping in developing effective DAA approaches for commercial UAVs. In particular, the LIDAR has been found to be more accurate at large distances from the platform (100–180 cm and up to 4 m), whereas the ToF sensor performed well at shorter ranges (0–120 cm). As shown in the work, the KF-based data fusion algorithm gave better results than Gelb's approach, at the cost of increased complexity. Nonetheless, the Kalman-based data fusion allows for easy real-time data processing and propagates the current state of knowledge of the dynamic measurements, upgrading the estimation error during the measurement process, a property extremely useful for statistical analysis and performance monitoring. Good performance of the preliminary system confirms the feasibility and robustness of this approach to an autonomous DAA system.

## References

1. Austin R (2010) Unmanned aircraft systems. Wiley, Hoboken, p 233
2. Valavanis KP (ed) (2008) Advances in unmanned aerial vehicles: state of the art and the road to autonomy, vol 33. Springer Science & Business Media, Berlin
3. Papa U, Del Core G, Giordano G, Ponte S (2017) Obstacle detection and ranging sensor integration for a small unmanned aircraft system. In: Proceedings of 2017 IEEE international workshop on metrology for aeroSpace (MetroAeroSpace). IEEE, pp 571–577
4. Ariante G, Papa U, Ponte S, Del Core G (2019) UAS for positioning and field mapping using LIDAR and IMU sensors data: Kalman filtering and integration. In: Proceedings of 2019 IEEE 5th international workshop on metrology for aerospace (metroaerospace). IEEE, pp 522–527
5. Papa U, Ariante G, Del Core G (2018) UAS aided landing and obstacle detection through LIDAR-sonar data. In: Proceedings of 2018 4th IEEE international workshop on metrology for aerospace. IEEE, pp 478–483
6. Papa U (2018) Embedded platforms for UAS landing path and obstacle detection. In: Studies in systems, decision and control, vol 136. Springer, Berlin
7. ATMEL (2014) ATMEL® ATmega2560/V 8-bit microcontroller with 64-KB in-system programmable flash—datasheet, Doc. 2549Q-AVR-02/2014, 435 p

8. Liggins ME, Hall DL, Llinas J (eds) (2009) Handbook of multisensor data fusion—theory and practice, 2nd edn. CRC Press, Taylor & Francis Group, Boca Raton, FL
9. Garmin (2016) Lidar Lite v3 operation manual and technical specifications. Garmin Ltd., Switzerland
10. STMicroelectronics NV (2016) World smallest time-of-flight ranging and gesture detection sensor, VL53L0X. Datasheet—production data
11. Gelb A (ed) (1974) Applied optimal estimation. MIT Press, Cambridge
12. Grewal MS, Andrews AP (2001) Kalman filtering—theory and practice using MATLAB. Wiley, Hoboken

# Design of Biofunctional Platforms: Differently Processed Biomaterials with Polydopamine Coating

**Simona Zuppolini**◉, **Iriczalli Cruz-Maya, Vincenzo Guarino**◉,
**Vincenzo Venditto, and Anna Borriello**◉

**Abstract** The chemical modification of biomaterials (i.e., polyesters, polysaccharides) can offer the opportunity to develop multifunctional platforms characterized by a combination of suitable properties and useful for biomedical applications. Polydopamine (PDA) is a mussel-inspired polymer with excellent properties such as adsorption ability, high hydrophilicity, electrical properties and numerous functional groups for immobilizing biomolecules. Due to high adhesion properties to almost all kinds of substrates PDA it has been largely used as coating for surface modification of material to modulate cellular responses, including cell spreading, migration, proliferation, and differentiation. In order to develop bio-functionalized systems, PDA was proposed as coating of biomaterials differently processed in form of sub-micrometric fibres, microgels and foams. Starting by a dopamine precursor solution, PDA coating deposition on substrates was performed by exploring different methodologies. In particular, poly-ε-caprolactone (PCL) electrospun fibres coated with PDA by electrofluidodynamic process was reported as possible bio-conductive interfaces. Preliminary results, in terms of synthesis conditions, structural characterization, and in vitro cell studies suggested that polymerization/oxidation reactions can be properly optimized to guide PDA self-assembly in order to control surface properties, and ultimately, influence cell materials interactions.

**Keywords** Polydopamine · Synthetic biomaterials · In vitro studies ·
Biofunctional surfaces

---

Iriczalli Cruz-Maya: Equal contribution.

---

S. Zuppolini (✉) · I. Cruz-Maya · V. Guarino · A. Borriello
Institute for Polymers, Composites and Biomaterials (IPCB), National Research Council, P.le E. Fermi, 80055 Portici, Naples, Italy
e-mail: simona.zuppolini@cnr.it

V. Venditto
Department of Chemistry, University of Salerno, Via S. Allende, 84081 Baronissi, SA, Italy

© The Author(s), under exclusive license to Springer Nature Switzerland AG 2021          17
G. Di Francia and C. Di Natale (eds.), *Sensors and Microsystems*,
Lecture Notes in Electrical Engineering 753,
https://doi.org/10.1007/978-3-030-69551-4_3

# 1 Introduction

Biomaterials have been conventionally defined as able to interact with biological systems by at scale ranging from macro-, micro- and nanoscale [1]. The interactive side with cells and tissues is generally the interface. Thus, the surface properties of biomaterials, including both natural and synthetic materials, are responsible for regulating biological responses that directly relate to clinical performance. The design and development of biocompatible materials achieving specific functionalities still represents an attractive challenge [2].

Surface engineering strategies have been exploited to tailor the physical and chemical properties of biomaterials with the aim of improving cellular interactions at the interface between biomaterials and physiological surroundings [3]. Multifunctional biomaterials can be also produced by combining amazing abilities in a wide range of methods and techniques including electrospinning, 3D printing, molecule self-assembly, and surface functionalization, and are envisaged for different biomedical applications including tissue engineering, medicine repair, drug delivery, medical devices. Surface functionalization of biomaterials via molecular design is a key approach to incorporate new tailored functionalities.

Polydopamine (PDA), a mussel-inspired polymer, perfectly fits to this aim, due to its excellent properties such as adsorption ability, high hydrophilicity, good chemical reactivity and high adhesion properties to almost all kinds of substrates [4, 5]. PDA structure consists of numerous functional groups, including amine and catechol moieties, providing a high versatility in surface chemistry and capability of immobilizing biomolecules [6]. Reductive and electrical properties increase the appealing of PDA which has been yet largely used on biomaterial surface to modulate cellular responses, including cell spreading, migration, proliferation, and differentiation [7]. PDA has been successfully coated on biomaterial structures as interface in order to enhance the in vivo functionality of biomedical implants [7, 8].

All these peculiar properties make this bioinspired polymer suitable for chemical modification of biomaterials (i.e., polyesters, polysaccharides) exploiting the opportunity to develop bio-systems with a combination of suitable properties.

As consequence, PDA can be proposed as coating of differently processed biomaterials (i.e., sub-micrometric fibres, microgels and foams) in order to develop multifunctional platforms for biomedical applications. Starting by a dopamine precursor solution, PDA coating deposition on substrates can be performed by exploring different methodologies under mild and versatile conditions to tune size, shape and morphology of thin films.

In this paper, we focused on poly-ε-caprolactone (PCL) electrospun fibres coated with PDA by electrofluidodynamic process and proposed as possible bio-conductive interfaces. Basing on our previous work [9] the aim was to explore the effect of different coating deposition of PDA on the fibre surface and cells interaction. Preliminary experiments, in terms of synthesis conditions, structural characterization, and in vitro cell suggested the opportunity to guide PDA self-assembly in order to control surface properties, and ultimately, influence cell materials interactions.

## 2 Experimental

### 2.1 Fabrication of PDA Coated PCL Electrospun Fibres

*Materials.* Poly-ε-caprolactone (PCL, $M_w$ 45 kDa) in pellet form (33% w/v), chloroform, sodium hydroxide (NaOH), dopamine hydrochloride (DA), ethanol and 2-amino-2-(hydroxymethyl)-1,3-propanediol (Tris base) were provided by Sigma Aldrich. All chemicals were of analytical grade and used as received.

*Electrospun fibres.* PCL fibres were fabricated by using a commercialized electrospinning equipment (NF500—MECC, Japan) following a procedure reported in a previous work [9]. Briefly, PCL ($M_w$ 65 kDa) in pellet form were dissolved in chloroform (33% w/v) and electrospun by using 16 kV as the voltage, 0.5 ml/h as the flow rate and 150 mm as electrode distance. In order to improve the hydrophilicity of electrospun fibres surface, an alkaline treatment was performed in NaOH solution (5 M) at room temperature, for 24 h.

*PDA deposition.* PDA coating was performed by a drop-to-drop deposition of a Tris-HCl 10 mM (pH 8.5) solution containing dopamine-hydrochloride (2 mg/mL) onto NaOH-treated PCL fibres. Deposition was performed at the beginning ($T_0$) or after completion of polymerization ($T_{24h}$), respectively.

### 2.2 Characterization of PDA Coated PCL Electrospun Fibres

Attenuated Total Reflectance-Fourier Transform Infrared (ATR-FTIR) spectra were recorded on a Perkin Elmer Spectrum 100 FTIR spectrophotometer in the 4000–400 $cm^{-1}$ region.

The morphology of coated electrospun fibres was investigated by Scanning Electron Microscopy images (SEM, Quanta FEG 200 FEI, The Netherlands) with the support of image analysis for fibre size quantification.

### 2.3 Biocompatibility Tests

In vitro studies were performed using human Mesenchymal Stem Cells (hMSC) at $5 \times 10^4$ onto nanofibres in Eagle's alpha minimum essential medium supplemented with 10% of fetal bovine serum, antibiotic solution (100 μg/ml streptomycin and 100 U/mL) and L-glutamine. Cell adhesion was assessed via crystal violet assay at 4 and 24 h of culture in triplicate while viability until 14 days via Cell Counting Kit-8 (CCK-8 kit) for cell counting.

## 3   Results and Discussion

A scheme of the entire process for PDA coated PCL electruspun fabrication is reported in Fig. 1. Hence, deposition on PCL fibres of dopamine precursor both at initial oxidative conditions ($T_0$) and at occurred polymerization to PDA ($T_{24h}$), were compared. In both cases, a changing to dark brown of surface substrates was observed thus confirming the PDA coating/deposition onto substrates.

The presence of PDA on the PCL surface was investigated by comparing ATR-IR spectra of coated and uncoated PCL electrospun fibres (Fig. 2). In all spectra, the characteristic peaks of PCL at 2950 and 2820 $cm^{-1}$ are predominant and attributable to $\nu$(C–H) stretching, while $\nu$(C=O) stretching was observed at 1725 $cm^{-1}$. After NaOH treatment, the hydrophilicity induced was confirmed by the broadband in 3600–3200 $cm^{-1}$ range associated to –OH groups confirming. Then, the presence of PDA on PCL surface is only slightly observed as expected due to low amount of PDA. However, in both $T_0$ and $T_{24h}$ cases, new weak peaks at 3186 and 1631 $cm^{-1}$ are attributable to $\nu$(N–H) of free amines and the aromatic C=C bonds in indole, respectively. The key absorption of catechol groups, and $\nu$(O–H), are included in 3600–3100 $cm^{-1}$ range.

Morphology of PCL surfaces before and after different treatments was observed by SEM images (Fig. 3) which clearly show a particle coating structure of PDA due to the deposition mechanism. At $T_0$, PDA coating appears uniformly distributed while pre-synthesized PDA deposition ($T_{24h}$) promotes the formation of sub-microsized clustered islands onto PCL fibres, that contribute to influence the topographic features (i.e., roughness) basically imparted to the fibre surface via alkaline pre-treatments [9]. In vitro experiments were performed to evaluate the effect of PDA treatment on hMSCs adhesion (Fig. 4a) and vitality (Fig. 4b). Firstly, the PDA-coated fibres showed that PDA surface treatment enhanced the cell adhesion at 24 h, respect to the PCL fibres, due to the hydrophilic properties of PDA. This is confirmed by optical images (Fig. 4a) that highlights a good cell spreading along the fibres, which is important for the initial cell-material interaction and the subsequently cellular behavior [10].

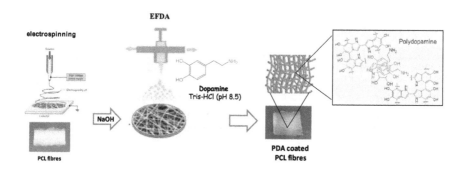

**Fig. 1** Schematic illustration of PDA coated PCL electrospun fibres preparation

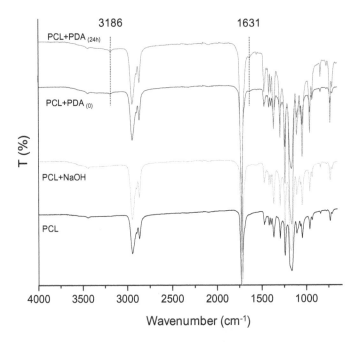

**Fig. 2** ATR-IR spectra of PCL electrospun fibres: untreated (black curve), NaOH treated (red curve), PDA coated at $T_0$ (blue curve) and $T_{24h}$ (magenta curve)

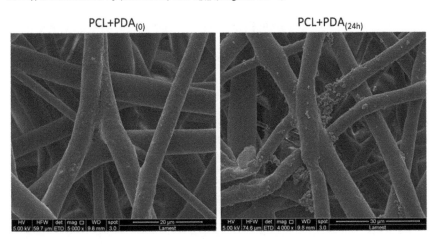

**Fig. 3** SEM images of PCL electrospun fibres with PDA coating deposited at $T_0$ (PCL + PDA$_0$) and $T_{24h}$ (PCL + PDA$_{24h}$)

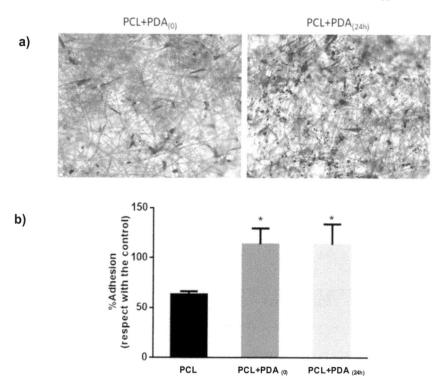

**Fig. 4** Cell adhesion of hMSC onto treated PCL fibres: **a** optical images after violet staining of cells and **b** quantitative analysis with CCK-8 assay (*significant difference against untreated PCL fibre; $p < 0.05$)

No statistical differences were recognized as a function of the PDA treatment.

Lastly, hMSCs viability was evaluated in all groups by CCK-8 kit. The formazan dye generated by the activity of dehydrogenases in cells is directly proportional to the living cells on fibre. According with these results, the presence of PDA is not affecting the cell viability at longer times (data not shown).

## 4 Conclusions

In this work, PCL fibrous platforms were fabricated and coated by PDA exploring different deposition conditions to drive the self-assembly of polymer on the surface. Morphological analyses confirmed that PDA can be synthesized and differently processed to form different surface coatings. Biocompatibility tests confirm a beneficial effect of PDA coating on in vitro cell response, also suggesting a slight effect due to different coating morphological properties.

In perspective, electrical conductivity of these functionalized surfaces will be evaluated to address their use as bio-conductive surface able to improve the in vitro interface with cell in pseudo 3D models.

# References

1. Jo YK, Kim HJ, Jeong Y, Joo KI, Cha HJ (2018) Biomimetic surface engineering of biomaterials by using recombinant mussel adhesive proteins. Adv Mater Interfaces 5:180006. https://doi.org/10.1002/admi.201800068
2. Perikamana SKM, Lee J, Lee YB, Shin YM, Lee EJ, Mikos AG, Shin H (2015) Materials from mussel-inspired chemistry for cell and tissue engineering applications. Biomacromolecules 16(9):2541–2555. https://doi.org/10.1021/acs.biomac.5b00852
3. Neto AI, Cibrão AC, Correia CR, Carvalho R, Luz GM, Ferrer GG, Botelho G, Picart C, Alves NM, Mano JF (2014) Nanostructured polymeric coatings based on chitosan and dopamine-modified hyaluronic acid for biomedical applications. Small 10(12):2459–2469. https://doi.org/10.1002/smll.201303568
4. Liebscher J (2019) Chemistry of polydopamine—scope, variation, and limitation. Eur J Org Chem 4976–4994. https://doi.org/10.1002/ejoc.201900445
5. Dreyer DR, Miller DJ, Freeman BD, Paul DR, Bielawski CW (2012) Elucidating the structure of poly(dopamine). Langmuir 28:6428–6435. https://doi.org/10.1021/la204831b
6. Cheng W, Zeng X, Chen H, Li Z, Zeng W, Mei L, Zhao Y (2019) Versatile polydopamine platforms: synthesis and promising applications for surface modification and advanced nanomedicine. ACS Nano 13:8537–8565. https://doi.org/10.1021/acsnano.9b04436
7. Lynge ME, van der Westen R, Postma A, Städler B (2011) Polydopamine—a nature-inspired polymer coating for biomedical science. Nanoscale 3:4916–4128. https://doi.org/10.1039/c1nr10969c
8. Lee H, Dellatore SM, Miller WM, Messersmith PB (2007) Mussel-inspired surface chemistry for multifunctional coatings. Science 318(5849):426–430. https://doi.org/10.1126/science.1147241
9. Zuppolini S, Cruz-Maya I, Guarino V, Borriello A (2020) Optimization of polydopamine coatings onto poly-ε-caprolactone electrospun fibres for the fabrication of bio-electroconductive interfaces. J Funct Biomater 11:19. https://doi.org/10.3390/jfb11010019
10. Wang H, Lin C, Zhang X, Lin K, Wang X, Shen SG (2019) Mussel-inspired polydopamine coating: a general strategy to enhance osteogenic differentiation and osseointegration for diverse implants. ACS Appl Mater Interfaces 11:7615–7625. https://doi.org/10.1021/acsami.8b21558

# Improvement of Ferroelectrics Properties of Lead-Free Thin Films by Sol Gel Process Optimization

V. Casuscelli, R. Scaldaferri, P. Aprea, P. S. Barbato, and D. Caputo

**Abstract** This work deals with the optimization of lead-free piezoelectric/ferroelectrics oxides thin films fabrication process. Such films, with formula $0.5(BaZr_{0.2}Ti_{0.8}O_3)-0.5(Ba_{0.7}Ca_{0.3}O_3)$ (BZT-BCT), were deposited on $Pt/TiO_x/SiO_2/Si$ wafers. The thin films fabrication process by sol-gel process consists of repeated chemical solution deposition routes, whose spin coating and heating parameters were optimized to reduce the residual stresses that are the major cause of cracking in the films. The results indicated that 250 nm thick films were obtained with limited presence of cracks and improved ferroelectric properties because of the enhanced morphological properties of the deposited layers.

**Keywords** Sol-gel thin films · BZT-BCT · Lead-free piezoelectric materials · Ferroelectric

## 1 Introduction

Barium Zirconate Titanate-Barium Calcium Titanate (BZT-BCT) is a promising piezoelectric/ferroelectric ceramic system [1–4] which can play an important role in developing capacitors, sensors, actuators and nonvolatile ferroelectric memory devices as an alternative to Lead Zirconate Titanate (PZT), currently the most used piezoelectric material in bulk or thin film form [4]. The growing concern about the potential impact on environment and human health related to the production and disposal of lead-based electronic devices fosters the research for a substitute material with comparable or even better properties. Moreover the development of piezoelectric micro-system devices typically requires the deposition of high-performance thin films on semiconductor substrates [5] and BZT-BCT could be considered one of

V. Casuscelli (✉) · R. Scaldaferri
Analog, MEMS and Sensors Group STMicroelectronics, Via Remo De Feo, 1, 80022 Arzano, Italy
e-mail: Valeria.Casuscelli@st.com

P. Aprea · P. S. Barbato · D. Caputo
Dipartimento di ingegneria Chimica, dei Materiali e della Produzione Industriale, Piazzale Vincenzo Tecchio, 80, 80125 Napoli, Italy

© The Author(s), under exclusive license to Springer Nature Switzerland AG 2021
G. Di Francia and C. Di Natale (eds.), *Sensors and Microsystems*,
Lecture Notes in Electrical Engineering 753,
https://doi.org/10.1007/978-3-030-69551-4_4

25

the best candidates to produce lead-free thin films with strong piezoelectric effect [6]. Chemical solution deposition techniques are widely used for the formation of binary/ternary oxide thin films thanks to the good stoichiometric control and to the possibility to induce crystallization at quite low temperatures with respect to other conventional methods [7, 8]. Spin coating is a frequently used deposition technique in the semiconductor industry because of the several advantages it offers, such as low cost, accurate control of the precursors chemistry, possibility to cover large surfaces and good control on film thickness and morphology. This method can also potentially guarantee the highest ferroelectric and piezoelectric properties of the final thin layer.

On the other hand, the outcomes of this process are strongly influenced by a huge number of parameters, which can be summarily described as (i) those related to the precursor solution production and (ii) those related to the deposition itself, like spin coating speed and time, pyrolysis and annealing temperature [9]. Each of them contributes to the morphology of the final layer, and the appearance of residual stresses during the deposition process from different contributions [9] is responsible for the presence of observable cracks on the surface of the obtained films. In this work, an analysis of the effect of the deposition parameters on the film properties has been performed, focusing on the spin coating speed and pyrolysis temperature, which influence the thickness of each deposited layer, the formation of cracks and the electric performances of BZT-BCT thin layer.

## 2 Experimental and Results

### 2.1 Materials and Methods

Thin films made of BZT-BCT were produced by depositing a precursor solution prepared as reported in [7] in a dry atmosphere with a final molarity of 0.25. The substrate used for the deposition was (111)-Pt/TiO$_2$/SiO$_2$/Si-(100). The electrical conductivity of platinum allows to use the substrate as a bottom electrode for piezo-electric applications. In addition, platinum is suitable for high temperature heat treatments, because of its negligible interaction with the film. To remove potential contaminants on the platinum surface the substrates were cleaned under a high-pressure nitrogen flow before the deposition. The precursor solution was deposited on the substrate by dispensing it in a spin coating system (Specialty Coating Systems model P-6712), according to a three-stage spinning program. Three different spin coating recipes have been chosen, varying the maximum velocity of the spinner for each stage, as reported in Table 1.

After the deposition of a solution layer, the desired ceramic phase was obtained by means of a sequence of three different heat treatments, the first two performed on a hot plate (ATV technologie GMBH HT-306), and the third into an industrial Rapid Temperature Processing machine (RTP, Steag 2800).

In detail, the whole standard process can be sketched as follows:

**Table 1** Used spin coating recipes

|  | A | B | C |
|---|---|---|---|
| First stage | 1600 rpm, 10 s | 2000 rpm, 10 s | 2800 rpm, 10 s |
| Second stage | 1800 rpm, 30 s | 2200 rpm, 30 s | 3000 rpm, 30 s |
| Third stage | 750 rpm, 10 s | 750 rpm, 10 s | 750 rpm, 10 s |

1. spin coating of the solution on the substrate
2. drying on hot plate at 150 °C for 5 min to achieve the evaporation of solvent
3. firing on hot plate at 450 °C for 5 min, to eliminate the residual solvent and burn off the organic fraction, obtaining an amorphous carbonate solid
4. crystallization in RTP at 830 °C for 60 s under pure oxygen flow to obtain the perovskite phase and achieve the final densification of the material.

Basing on this general scheme, two options have been investigated: changing the pyrolysis temperature (375 °C) and adding a third thermal treatment at the same temperature of the first one to smooth the cooling rate of the amorphous gel.

After the first cycle, three more layers of BZT-BCT where deposited on the same substrate repeating the sequence 1–3 for each layer, before treating the sample again in RTP.

BZT–0.5BCT thin films were characterized in capacitors structures with platinum electrodes. Devices were realized by sputtering top electrodes through a shadow mask onto the film surfaces, to produce a final stack $Si/SiO_2/TiO_2/Pt$ (100 nm)/BZT-0.5BCT thin film/Pt (98 nm). Ferroelectric characteristics of such devices have been measured at room temperature by Thin Film Analyzer (TFA2000 Aixact) as showed in Fig. 1. By this measurement polarization fields loops have been acquired to evaluate the ferroelectric performances.

According to the standard definitions of Remanent Polarization (Pr) and coercive Field (Ec) [10] $\Delta Pr$ can be estimated from ferroelectric curves as the difference between the values of Remanent polarization at zero Electric Field and $\Delta Ec$ can be

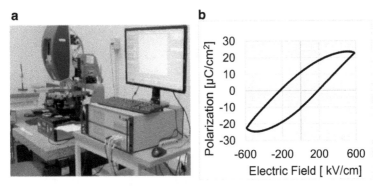

**Fig. 1** **a** TFA2000 Aixact equipment, **b** typical ferroelectric characteristic

evaluated as the difference between the values of Electric Field corresponding to a null Remanent polarization.

In the following the characterization data for a $1 + 3 + 3$ (7 layers deposition) sample (i.e. with crystallization step after the first, the fourth and the seventh layers deposition) are reported.

## 2.2 Results

The results of ferroelectric characterization of the 7-layer thin films obtained at different spin coating velocities (A, B and C) are reported in Fig. 2 and Table 2.

Regarding the values of the coercive fields, a significant and continuous increase emerged from the samples $1 + 6$ A 150–450, $1 + 6$ B 150–450 and $1 + 6$ C 150–450 (76.94 kV/cm, 178.6 kV/cm and 310.82 kV/cm, respectively) which is related to the progressive decreasing size of the film grains and thickness.

Then starting from the C recipe different treatment temperatures have been investigated.

Comparing ferroelectric parameters for the samples obtained by using the same spin coating speed but different treatment temperatures, the results seems to be very different (see Fig. 3).

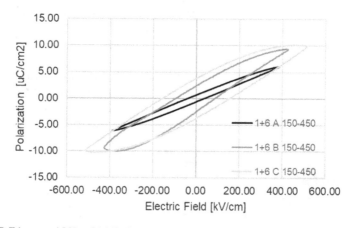

**Fig. 2** P-E loops at 15 V and 1 kHz for sample produced by different spin coating recipe

**Table 2** Coercive fields ($\Delta$Ec), remanent polarization ($\Delta$Pr) value obtained by P-E loops and dielectric constant ($\varepsilon$) obtained from CV diagrams for $1 + 6$ samples

| Sample | $\Delta$Ec [kV/cm] | $\Delta$Pr [$\mu$C/cm$^2$] |
|---|---|---|
| $1 + 6$ A 150–450 | 76.94 | 1.24 |
| $1 + 6$ B 150–450 | 178.60 | 4.90 |
| $1 + 6$ C 150–450 | 310.82 | 7.56 |

**Fig. 3** P-E loops measured for films obtained by different thermal treatments and measured at 1 kHz

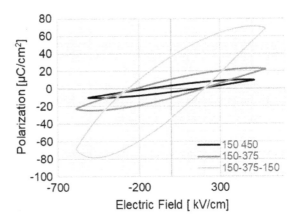

In Table 3 the estimation of $\Delta Ec$ and $\Delta Pr$ values obtained from PE curves of films have been reported.

It is worth noting that the $\Delta Ec$ values of the last samples (C recipes) are closer, since they have almost the same thickness close to 250 nm.

Moreover the direct comparison of data for thin films produced at the same spin coating velocity and different heat treatments indicate that the ferroelectric behavior of the thin films is better in the case of 1 + 6 C 150–375–150 sample in agreement with the improved morphology of this sample. In fact, SEM characterization of the films showed a regular morphology and a limited presence of cracks and microscopic defects as could be observed in Fig. 4.

**Table 3** Coercive fields ($\Delta Ec$) remanent polarization ($\Delta Pr$) value obtained by P-E loops for 1 + 6 samples using different temperature

| Sample | Temperatures [°C] | $\Delta Ec$ [kV/cm] | $\Delta Pr$ [$\mu C/cm^2$] |
|---|---|---|---|
| C 150–450 | 150–450 | 310.82 | 7.56 |
| C 150–375 | 150–375 | 344.35 | 17.82 |
| C 150–375–150 | 150–375–150 | 441.10 | 66.52 |

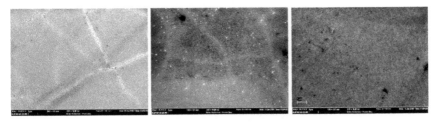

**Fig. 4** SEM micrographs (20KX) of 1 + 6 samples obtained at the same spin coating recipe C and different heat treatments 150–450, 150–375 and 150–375–150

# 3 Conclusion

Thin films of BZT-BCT perovskite piezoelectric oxide with a thickness of 250 nm have been successfully deposited by means of an optimized spin coating process onto a suitable substrate. SEM analysis showed that optimized recipe allows to obtain an improved morphology and ferroelectric analysis confirmed improved performances reasonably coherent with literature data.

# References

1. Damjanovic D, Biancoli A, Batooli L, Vahabzadeh A, Trodahl J (2012) Elastic, dielectric, and piezoelectric anomalies and Raman spectroscopy of $0.5Ba(Ti_{0.8}Zr_{0.2})O_3$–$0.5(Ba_{0.7}Ca_{0.3})TiO_3$. Appl Phys Lett 100:192907, 1–4
2. Adhikari P, Mazumder R, Sahoo GK (2016) Electrical and mechanical properties of $0.5Ba(Zr_{0.2}Ti_{0.8})O_3$–$0.5(Ba_{0.7}Ca_{0.3})TiO_3$ (BZT–BCT) lead free ferroelectric ceramics reinforced with nano-sized $Al_2O_3$. Ferroelectrics 490:60–69
3. Liu WF, Ren XB (2009) Large piezoelectric effect in Pb-free ceramics. Phys Rev Lett 103(25):257602-1-4
4. Panda PK, Sahoo B (2015) PZT to lead free piezoceramics: a review. Ferroelectrics 474:128–143
5. Luo BC, Wang DY, Duan MM, Li S (2013) Growth and characterization of lead-free piezoelectric $BaZr_{0.2}Ti_{0.8}O_3$–$Ba_{0.7}Ca_{0.3}TiO_3$ thin films on Si substrates. Appl Surf Sci 270:377–381
6. Lima EC, Araujo EB (2016) Role of residual stress on phase transformations of $Pb(Zr_{0.50}Ti_{0.50})O_3$ thin films obtained from chemical route. Ferroelectrics 499:28–35
7. Aprea P, Liguori B, Caputo D, Casuscelli V, Cimmino A, Di Matteo A, Salzillo G (2017) Green chemical routes for the synthesis of lead-free ferroelectric material $0.5Ba(Zr_{0.2}Ti_{0.8})O3$–$0.5(Ba_{0.7}Ca_{0.3})TiO_3$. Adv Sci Lett 23:6015–6019
8. Solayappan N, Joshi V, DeVilbiss A, Bacon J, Cuchiaro J, McMillan L, Paz de Araujo C (1998) Chemical solution deposition (CSD) and characterization of ferroelectric and dielectric thin films. Integr Ferroelectr 22(1–4):1–11
9. Kang G, Yao K, Wang J (2012) $(1-x)Ba(Zr_{0.2}Ti_{0.8})O_3$-$x(Ba_{0.7}Ca_{0.3})TiO_3$ ferroelectric thin films prepared from chemical solutions. J Am Ceram Soc 95(3):986–991
10. Damjanovic D (2005) In: Mayergoyz I, Bertotti G (eds) The science of hysteresis, vol 3. Elsevier, Amsterdam

# Lock-In Thermography to Visualize Optical Fibres Buried Inside 3D Printed PLA Items

**Simone Boccardi, Giuseppe del Core, Pasquale di Palma, Agostino Iadicicco, Stefania Campopiano, and Carosena Meola**

**Abstract** This paper shows some preliminary results of non-destructive tests performed by means of lock-in thermography on 3D printed items embedding optical fibres. The intention is to verify if lock-in thermography can be a valid means to identify the path of optical fibres which are buried inside a material and that cannot be identified with the naked eye. To this end feasibility tests have been carried out on some PLA samples manufactured by using the fused deposition modelling technique and by varying some key parameters.

**Keywords** Lock-in thermography · Optical-fibres · Non-destructive testing · 3D printing

## 1 Introduction

Currently the optical fibres are used in a wide range of applications such as civil, marine, mechanical, military, power generation, offshore and oil and gas [1]. Moreover, they seem to be very well suited for online structural health monitoring of composite materials especially for aerospace applications. In fact, optical fibres can be easily embedded inside composite laminates thanks to their small size, low weight, good mechanical resistance, flexibility and resistance to heat. For these reasons optical fibres are ever more attracting the attention of the industrial and academic communities as regards the development of new methodologies and applications for continuous structural health monitoring such as strain measurement and damage detection [2, 3]. Surely, the reliability of measurements performed by means of optical fibres depends also on their proper positioning inside the material. Displacements and

S. Boccardi (✉) · G. del Core · P. di Palma · A. Iadicicco · S. Campopiano
Dipartimento di Scienze e Tecnologie e Dipartimento di Ingegneria, Centro Direzionale,
Università degli Studi di Napoli Parthenope, isola C4, 80143 Napoli, Italy
e-mail: simone.boccardi@uniparthenope.it

C. Meola
Dipartimento di Ingegneria Industriale, Università degli Studi di Napoli Federico II, Via Claudio
21, 80125 Napoli, Italy

G. Di Francia and C. Di Natale (eds.), *Sensors and Microsystems*,
Lecture Notes in Electrical Engineering 753,
https://doi.org/10.1007/978-3-030-69551-4_5

deviations from the desired position may occur during manufacturing or operating life and can significantly affect measurements by introducing a significative noise on the acquired data. In this context infrared thermography [4] can be considered a valid means to individuate the path of the hidden optical fibres in the final product. In particular, lock-in thermography (LT) has proved its effectiveness to detect many types of defects and damages in composite materials such as delamination, inhomogeneity in matrix distribution, porosity and impact damage [5]. Of course, the minimum dimension of the detectable stuff hidden inside a material, depends on many factors that characterize the inclusion, such as dimensions, depth and thermal properties with respect to the hosting material [4]. Some preliminary tests have underlined the effectiveness of LT in the detection of artificial and production defects in PLA specimens manufactured by means of fused deposition modelling (FDM) [6]. The idea behind this paper is to understand if LT can be exploited to locate the position of optical fibres inside the material and obtain other useful information about them. For this investigation it was chosen to manufacture PLA specimens by means of the 3D printing FDM for several reasons; it is a relative easy and cheap method, it allows to easily positioning the optical fibres at a given depth during production, it allows for visual inspection of all the production steps to get information about likely occurrence of defects during manufacturing.

## 2 Experimental and Results

### 2.1 Materials and Samples

Two square specimens, which are named A and B, were manufactured by means of the fused deposition modelling technique (FDM) by using Polylactic Acid (PLA) by Renkforce as raw thermoplastic material.

The production parameters used during manufacturing involve: an extrusion temperature of 205 °C, an infill section of 100% and a nominal layer thickness of 0.2 mm. Classical optical fibres Corning SMF-28 with core, cladding and coating diameters of 8, 125 and 250 μm respectively, were properly embedded inside the specimens. In some cases, the external protection coating was removed before to place the optical fibre inside the material in order to investigate the effects introduced by the external coating on the detectability by lock-in thermography. During the FDM 3D printing process several layers of PLA, 0.2 mm thick, were overlapped each other's at ±45° to obtain a specimen of desired thickness. Some details of the specimens are collected in Table 1.

To incorporate the optical fibres inside the samples, the 3D printing process was temporarily stopped, and the optical fibres were horizontally placed (0°) at about the same distance from each other's over the last printed layer (Fig. 1a). Then the 3D printing process starts buck again and other layers were printed over the optical fibres in order to incorporate them inside the material.

**Table 1** Specimen manufacturing details

| Sample | Thickness (mm) | Number of layers | Sample side (mm) | Fibres depth from S1 | Fibres depth from S2 |
|--------|---------------|------------------|------------------|----------------------|----------------------|
| A | 0.6 | 3 | 100 | 2 layers, 0.4 mm | 1 layer, 0.2 mm |
| B | 1.4 | 7 | 80 | 4 layers, 0.8 mm | 3 layers, 0.6 mm |

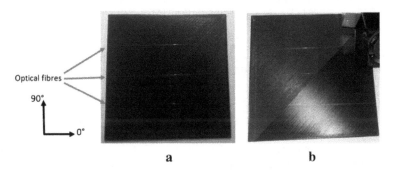

a         b

**Fig. 1** Photos of specimen A taken during the production phase. **a** Optical fibres placement, **b** optical fibres partially covered while a new layer is just being printed over them

The symbols S1 and S2 are used to indicate the specimen upper and the lower surfaces respectively. S2 is the surface in contact with the printing plate during the 3D printing process.

The optical fibres are placed closer to the S2 surface in order to investigate the fibre's detectability in function of depth from the surface by limiting the number of specimens required (see Table 1). In Figs. 2 and 3 are shown same images of the upper surfaces S1 for the A and B specimens respectively that illustrate some detail about the fibres positioning. In relation with the specimen A, the optical fibres on top and on bottom are fully covered by their protection coating whilst the optical fibre in the middle is only half covered by the protection coating (Fig. 2c). In specimen B only the optical fibre on the top is fully coated, that on the bottom is completely bare,

**Fig. 2** Specimen A details; **a** scheme of the fibre type and positioning through the thickness, **b** photo of S1 surface, **c** scheme of the specimen plan view with fibres details and positioning

**Fig. 3** Specimen B details; **a** scheme of the fibre type and positioning through the thickness, **b** S1 surface photo, **c** scheme of the specimen plan view with fibres details and positioning

whilst on a par with the specimen A, the protection coating is only half removed in the optical fibre placed in the middle (Fig. 3c).

## 2.2 Test-Setup

Each specimen was inspected by means of LT in reflection [7] from both surfaces. Two halogen lamps, of 1000 W each, thermally stimulates the specimen with a heat flux harmonically modulated at selectable values of frequency $f$ by means of the lock-in module. During the harmonic heating the infrared camera Flir SC6000, equipped with QWIP detector, records a sequence of images which are analysed by means of the IRLock-In© software. Results are presented in terms of phase images containing information about the material conditions. The used test setup, including the infrared camera, the halogen lamps and the test specimen, is shown in Fig. 4.

LT performed in reflection allows to inspect the material at several depths p form the observed surface, by changing the heating frequency $f$, according with the Eq. 1:

**Fig. 4** Test setup

$$p \approx 1.8\sqrt{\frac{\alpha}{\pi f}} \qquad (1)$$

where $\alpha$ is the material thermal diffusivity. If $\alpha$ is known it is possible to evaluate the depth of inspection p, otherwise by lowering $f$ it is only possible to obtain phase images that are representative of a relatively higher value of depth p with respect to the observed surface. Since feasibility tests are performed, the $\alpha$ value is not evaluated and only qualitative results are shown. For each specimen, several tests have been performed at different values of the heating frequency (from 0.6 down to 0.02 Hz) in order to inspect the specimen through the thickness starting from the observed surface up to intercept the optical fibres.

## 2.3   Results and Discussion

Some results in terms of phase images, relative to LT tests performed on the surface S1 of both specimens A and B for several values of the heating frequency $f$, are shown in Figs. 5 and 6. As general observation, in both specimens it is possible

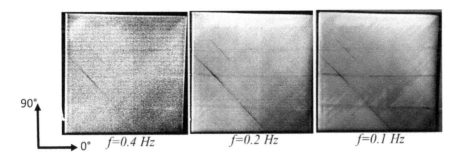

**Fig. 5**  Phase images of specimen A taken from surface S1; **a** $f = 0.4$ Hz, **b** $f = 0.2$ Hz, **c** $f = 0.1$ Hz

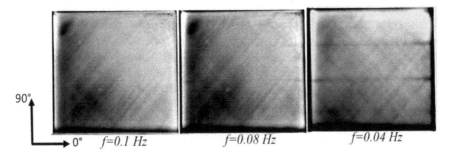

**Fig. 6**  Phase images of specimen B taken from surface S1; **a** $f = 0.1$ Hz, **b** $f = 0.08$ Hz, **c** $f = 0.04$ Hz

to clearly recognize the optical fibres buried inside the material (see also Figs. 2 and 3), the printing directions pattern of ±45° and also some defects as well non uniform regions due to production problems. It is possible to recognize the following main manufacturing defects: a wide non uniform area at the centre of the specimen A, which modifies going in depth (Fig. 5a–c), two dark strips at −45°, that can be seen also at naked eye over the surface S1 of specimen A (Figs. 5a–c, 2a), a dark non uniform region in the top left corner of specimen B (Fig. 6a, b) and a dark non uniform wide area in the bottom left region of specimen B (Fig. 6a, b).

In both the specimens A and B the optical fibres become clearly visible at the lower value of the heating frequency (Figs. 5c and 6c), when the depth of inspection entirely overcomes the layer where the optical fibres are placed. In specimen A the optical fibres appear clearly visible for higher values of $f$ ($f = 0.1$ Hz) because they are closer to the S1 surface with respect to fibres in specimen B ($f = 0.04$ Hz) (see Table 1 and Eq. 1). Besides, the optical fibres in specimen B appear more blurred with respect to those inside specimen A; this because of the lateral heat diffusion through the thickness [7].

In specimen A the coated fibres (on top and bottom) appear differently contoured whilst the half-coated fibre appears more countered along the bare tract. Those unexpected results are probably due to the interaction of the optical fibres with the more uniform neighbouring area.

In specimen B the upper fully coated fibre and the middle half coated fibre appear more countered with respect to the fully bare fibre (Fig. 6a–c) on the bottom, that is only barely visible. In order to confirm the observed results and exclude any secondary effects, some tests were repeated at the same values of the heating frequency by placing the specimens upside down; the same results were obtained.

## 3  Conclusion

Some specimens, manufactured by means of 3D printing FDM process and with embedded optical fibres, have been non-destructively evaluated with lock-in thermography. The heating frequency $f$ was varied to inspect the specimens at several depths through the thickness in order to detect the embedded optical fibres. The obtained results seem promising for the considered material and thickness. In fact, it was always possible to identify the optical fibres buried inside the material and get also information, in terms of phase differences, about the presence of the optical fibres protection coating and depth from the surface at which the optical fibres are located. In addition, it was possible to underline the presence of some defects and non-uniformity due to manufacturing problems.

Of course, this is only a preliminary investigation and other tests should be performed by considering other materials and production methods in order to clearly asses the LT potentialities and limitations in the detection of very thin buried optical fibres.

**Acknowledgements** The authors thanks Dr. Salvatore Barile for the provided technical support.

# References

1. Saeter E, Lasn K, Echtermeyer AT et al (2019) Embedded optical fibres for monitoring pressurization and impact of filament wound cylinders. Compos Struct 210:608–617
2. Minakuchi S, Takeda N (2013) Recent advancement in optical fiber sensing for aerospace composite structures. Photonic Sens 3(4):345–354
3. Di Palma P, Palumbo G et al (2019) Deflection monitoring of bi-dimensional structures by fiber Bragg gratings strain sensors. IEEE Sens J 19(11)
4. Vollmer M, Möllmann K (2018) Infrared thermal imaging. In: Fundamentals, research and applications, 2nd edn. Wiley. ISBN: 978-3-527-41351-5
5. Ripley B (2001) The R project in statistical computing. MSOR Connections Newsl LTSN Maths Stats OR Netw 1.1:23–25
6. Boccardi S, Carlomagno GM et al (2019) Lock-in thermography for non-destructive testing of 3D printed PLA items. In: Lecture notes in electrical engineering, vol 629, pp 149–155
7. Meola C, Boccardi S, Carlomagno GM (2016) Infrared thermography in the evaluation of aerospace composite materials. Woodhead Publishing. Print Book ISBN: 9781782421719, 180 p

# Disease Biomarkers Detection in Breath with a Miniaturized Electronic Nose

J. P. Santos⊙, C. Sanchez-Vicente⊙, J. Lozano⊙, F. Meléndez⊙,
P. Arroyo⊙, and J. I. Suarez⊙

**Abstract** A miniaturized wireless electronic nose for the detection of diseases through the breath is introduced in this communication. The device presented has an electronic design similar to previous prototypes, but with a smaller size and consumption. It is equipped with four miniaturized digital gas sensors and an integrated temperature and humidity sensor. Power is provided by a battery that can be charged through a micro USB connector. The system is connected through a low-energy Bluetooth connection. Different biomarkers corresponding to three diseases have been selected to verify the capacity of the prototype to discriminate among them. CO, NO and $NO_2$ have been selected as biomarkers of COPD or asthma, and acetone for diabetes. The e-nose has been used to measure the selected biomarkers at different concentrations, corresponding with different levels of gravity of illnesses. The target compound mixtures are generated from calibrated gas bullets and permeation tubes, with relative humidity up to 50%.

**Keywords** Electronic nose · Miniaturized · Low consumption · Gas sensors · e-nose · Biomarkers · Breath · Asthma · COPD · Diabetes · Volatile organic compounds · Chronic diseases · Digestive disease · Respiratory diseases

## 1 Introduction

At present, the environmental conditions in cities have generated an increase in the number of people with respiratory problems. This factor, together with a stressful lifestyle and poor nutrition has led to an increase in the percentage of patients with chronic diseases. These types of pathologies such as heart disease, cancer or respiratory disease, among others, are leading causes of death in the world, being responsible

J. P. Santos (✉) · C. Sanchez-Vicente
Institute of Physics Technology and Information, ITEFI-CSIC, Madrid, Spain
e-mail: jp.santos@csic.es

J. Lozano · F. Meléndez · P. Arroyo · J. I. Suarez
School of Industrial Engineering, University of Extremadura, Badajoz, Spain

© The Author(s), under exclusive license to Springer Nature Switzerland AG 2021
G. Di Francia and C. Di Natale (eds.), *Sensors and Microsystems*,
Lecture Notes in Electrical Engineering 753,
https://doi.org/10.1007/978-3-030-69551-4_6

for 63% of deaths according to the WHO. Some of these chronic diseases such as diabetes, asthma or COPD are given less importance because they do not have very high mortality rates [1, 2].

These affections have high morbidity, asthma and diabetes affect to 339 million and 422 million of patients respectively of the world. In recent years the number of patients affected by these pathologies has increased significantly, in the case of diabetes increased by 25% in 30 years. For this reason, the development of rapid diagnostic equipment is necessary to effectively control these chronic diseases. Therefore, incorrect diagnosis or inadequate control can cause an exacerbation of the disease. In addition, life expectancy (especially in developed countries) has increased in recent years, so this factor becomes more relevant for patients to have a better quality of life [1, 3].

Electronic nose technology allows non-invasive analyses and development of small and portable devices. This offers the possibility that the patient can realise the control analysis of the disease from home, and the data are automatically sent to the doctor through an app. In this way, the health system can also be relieved.

This type of technology not only contributes to improving the quality of life of patients but also has a positive influence on the health system. With this type of low-cost device, more people can be given access to disease control. In addition, a significant reduction in health costs arising from medical care and hospitalisations will be achieved.

## 2 Electronic Nose Prototype and App

The device presented has an electronic design similar to previous prototypes (Winose6) [4], but with a smaller size and consumption. This home-designed (UEx) device includes four miniaturized digital gas sensors (BME680, CCS811, iAQ-Core, and SGP30) and an integrated temperature and humidity sensor (SHT21). The system control is based on the use of a PIC32MM0256GPM048 microcontroller. The schematic electronic nose is shown in Fig. 1. Power is provided by a Li-Ion battery that can be charged through a micro USB connector included in the design.

The olfactory device is connected wirelessly through a low energy Bluetooth connection or by serial communication via UART bus. The communication protocol is based on the use of commands for the configuration of the system and the reception of measurements.

Besides the size and cost, the main difference between this electronic nose and others is the type of sensor used. In this case, digital gas sensors have been used that are capable of filtering and pre-processing the signal, offering a more robust signal than the analog sensors used in other devices.

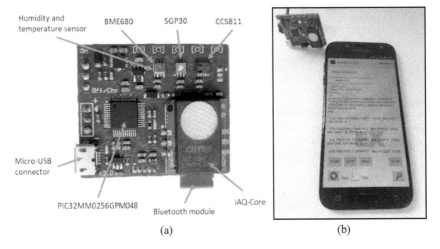

Fig. 1 **a** Schematic of electronic nose. **b** Mobile app and miniaturized e-nose

## 3 Biomarkers

The breath is a mixture of nitrogen, oxygen, carbon dioxide, water, and other gases at a lower concentrations. The air exhaled of a person has around 2500 volatile organic compounds (VOCs) with 95% relative humidity (RH). Acetone is a volatile organic compound related to diabetes. On the other hand, other gaseous compounds as carbon monoxide (CO), nitric oxide (NO) or nitrogen dioxide ($NO_2$) can be detected and evaluated as potential biomarkers of lung diseases as asthma or COPD [5–7].

To detect these diseases, different gas samples have been prepared with the selected biomarkers for these diseases. Table 1 shows the concentration ranges for healthy and sick people for each biomarker used.

$NO_x$ can be found in the literature instead of distinguishing between NO and $NO_2$. Because in some cases it may be difficult to determine exactly which compound is in the gas mixture. For this reason, it has been decided to use this terminology in Table 1.

**Table 1** Concentration range used for the different biomarkers

|         | Range for healthy persons | Range for ill persons |
|---------|---------------------------|-----------------------|
| Acetone | 0.5–2 ppm                 | >4 ppm                |
| $NO_x$  | 1–20 ppb                  | >25 ppb               |
| CO      | 1–2 ppm                   | >5 ppm                |

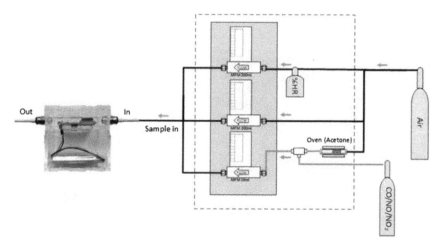

**Fig. 2** Measurement setup

## 4 Experimental Procedure

Different samples that simulate human breath have been prepared in the laboratory, using the measurement setup that it shows in Fig. 2. To generate the acetone
samples at different concentrations, a permeation tube has been used. On the other
hand, NO, $NO_2$ and CO gas bottles have been used to prepare the different samples
corresponding to asthmatics or COPD patients.

As stated above, the breath has around 95% of relative humidity. In this preliminary
work, a 50% of relative humidity value has been used instead. Measurements have
been made also in dry air in order to determine the influence of the humidity on the
sensors. A humidifier bottle has been used to obtain the samples with 50% of relative
humidity.

Besides, three mass flow controllers have been used to obtain the desired concentrations of each gas. Once, the gas mixture has been prepared, it is passed through
the sensor cell. Measurements have been made at different concentrations for each
of the gases selected as biomarkers. Each analysis has been done with and without
humidity. The operation temperature of the sensors was fixed around 300 °C.

## 5 Results

In Fig. 3 the responses of each sensor to the acetone, CO and $NO_x$ respectively are
shown. In addition, the result obtained for the sample without humidity and with a
relative humidity of 50% is included. The concentration range for healthy people is
indicated in green, while for ill people it is indicated in red. The response is indicated
as a percentage of the sensor resistance variation. It should be noted that the response

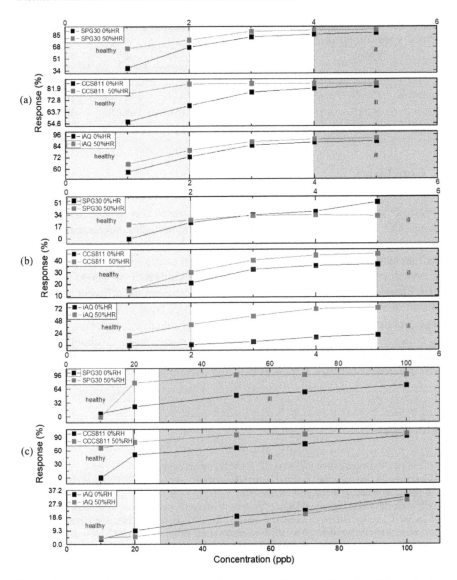

**Fig. 3** Representation of sensor responses for samples with and without humidity. **a** Acetone samples. **b** CO samples. **c** NO$_x$ samples

of the BME680 sensor has not been included in the graphs, because these compounds saturate the sensor and no distinction is made between the different concentrations. Table 2 shows the percentage variation between the minimum response value corresponding to a sick and the maximum response value corresponding to a healthy person for each gas.

**Table 2** Percentage variation between the minimum response in a sick person and maximum response in a healthy person, for each biomarker

|         | SPG30 | | CCS811 | | iAQ | |
|---------|--------|--------|--------|--------|--------|--------|
|         | 0% RH | 50% RH | 0% RH | 50% RH | 0% RH | 50% RH |
| Acetone | 18.36% | 14.23% | 13.28% | 0.17% | 14.90% | 11.48% |
| $NO_x$ | 25.72% | 18.49% | 15.43% | 15.65% | 09.87% | 08.47% |
| CO | 28.89% | 05.83% | 15.63% | 15.43% | 19.23% | 33.01% |

# 6 Conclusion

To evaluate the effectiveness of the equipment, measurements have been made with acetone, CO, $NO_x$ at different concentrations. In addition, these measurements have been made without humidity and with 50% relative humidity, to determine the influence that humidity has on the sensors. The different samples have been prepared in the laboratory using permeation tubes (acetone) or calibrated gas bottles (CO, NO, $NO_2$). The electronic nose is able to differentiate between the different concentrations of biomarkers corresponding to patients with diabetes or asthma and healthy people. As can be seen in Table 2, humidity generally attenuates the sensor response and causes an increase in the response compared to measurements without humidity. Humidity affects the SG30 sensor more than the iAQ sensor, for which there are fewer differences between responses with and without humidity.

# References

1. World Health Organization (WHO) (2017) Global report on diabetes
2. World Health Organization (WHO) (2016) Global report on diabetes WHO library cataloguing-in-publication data global report on diabetes
3. World Health Organization (WHO). https://www.who.int/topics/chronic_diseases/es/
4. Sánchez C, Pedro Santos J, Lozano J, Sayago I (2020) Hand-held electronic nose to detect biomarkers of diseases through breath. In: Sensors and microsystems. AISEM annual conference on sensor and microsystems. Springer, Berlin
5. Španěl P, Dryahina K, Rejšková A, Chippendale TWE, Smith D (2011) Breath acetone concentration. Biological variability and the influence of diet. Physiol Meas 32(8)
6. Sánchez C, Santos JP, Lozano J (2019) Use of electronic noses for diagnosis of digestive and respiratory diseases through the breath. Biosensors (2019)
7. Shorter JH, Nelson DD, McManus JB, Zahniser MS, Sama SR, Milton DK (2011) Clinical study of multiple breath biomarkers of asthma and COPD (NO, CO(2), CO and N(2)O) by infrared laser spectroscopy. J Breath Res 5(3):037108

# Vacuum Gauge from Ultrathin MoS$_2$ Transistor

**A. Di Bartolomeo, A. Pelella, A. Grillo, F. Urban, L. Iemmo, E. Faella, N. Martucciello, and F. Giubileo**

**Abstract** We fabricate monolayer MoS$_2$ field effect transistors and study their electrical characteristics from $10^{-6}$ Torr to atmospheric air pressure. We show that the threshold voltage increases for growing pressure. Hence, we propose the transistors as air pressure sensors, showing that they are suitable as low-power vacuum gauges. The devices operate on the pressure-dependent O$_2$, N$_2$ and H$_2$O molecule adsorption that affects the n-doping of the MoS$_2$ channel.

**Keywords** Molybdenum disulfide · 2D materials · Transistors · Pressure sensor · Adsorbates

## 1 Introduction

Following the great success of graphene [1–4], several families of atomically thin materials have emerged in the past decades and have been dominating the material research scenario [5, 6]. In particular, two-dimensional (2D) transition metal dichalcogenides (TMDs) have attracted a lot of attention due to several promising properties for electronic, optoelectronic, energy, catalysis and sensing applications [7–9]. TMDs consist of a "sandwich" layered structure with a transition-metal sheet located in between two chalcogen sheets and possess unique properties such as energy bandgap tunable by the number of layers (from 0 to about 2.2 eV), good mobility up to few hundreds cm$^2$V$^{-1}$ s$^{-1}$, photoluminescence, broadband light adsorption, surface without out-of-plane dangling bonds that allows the fabrication of heterostructures, high strength with Young's modulus up to 300 GPa, exceptional flexibility, and thermal stability in air [10–12]. They can be produced by mechanical or liquid

A. Di Bartolomeo (✉) · A. Pelella · A. Grillo · F. Urban · L. Iemmo · E. Faella
Dipartimento di Fisica "E.R. Caianiello", via Giovanni Paolo II, 132, 84084 Fisciano, Salerno, Italy
e-mail: adibartolomeo@unisa.it

A. Di Bartolomeo · A. Pelella · A. Grillo · F. Urban · L. Iemmo · E. Faella · N. Martucciello · F. Giubileo
CNR-Spin, via Giovanni Paolo II, 132, 84084 Fisciano, Salerno, Italy

© The Author(s), under exclusive license to Springer Nature Switzerland AG 2021    45
G. Di Francia and C. Di Natale (eds.), *Sensors and Microsystems*,
Lecture Notes in Electrical Engineering 753,
https://doi.org/10.1007/978-3-030-69551-4_7

exfoliation, chemical vapor deposition (CVD), molecular beam epitaxy, pulsed laser deposition, etc. [13].

Molybdenum disulfide ($MoS_2$) is formed by covalently bonded S–Mo–S sequences held together by weak van der Waals forces, resulting in easy-to-exfoliate 2D layers [14, 15]. $MoS_2$ is a semiconductor with 1.2 eV indirect bandgap in the bulk form that widens up to 1.8–1.9 eV and becomes direct in the monolayer. It is a promising material for field-effect transistors (FETs) with high performance and on/off ratio [16–18], sensitive broadband photodetectors [19, 20], catalysis [21], chemical and biological [22–25] or strain and pressure sensors [23, 26].

Microscopic pressure sensors that can rapidly detect small pressure variations are of high demand in robotic technologies, human–machine interfaces, electronic skin, sound wave detection, and health monitoring devices. Pressure sensors are very important in many other fields, such as automobiles, aircrafts, well drilling, and medical applications.

The exceptional mechanical properties of $MoS_2$ nanosheets [27] have inspired their application as ultrathin diaphragms capable of large deflection deformations at low pressure to achieve high sensitivity in pressure sensors. For instance, a thin and sensitive diaphragm is attached onto one end face of a cleaved optical fiber to form an extrinsic Fabry–Perot interferometric structure that detects the applied pressure through the measurement of the deflection deformation of the diaphragm. Fabry–Perot ultrasensitive pressure sensors with nearly synchronous pressure–deflection responses have been fabricated using few-layer $MoS_2$ films. Compared to conventional diaphragm materials (e.g., silica, silver films), they have allowed to achieve three orders of magnitude higher sensitivity (89.3 nm $Pa^{-1}$) [28].

Highly sensitive pressure sensors have been fabricated by integrating a conductive microstructured air-gap gate with $MoS_2$ transistors. The air-gap gate is used as the pressure-sensitive gate for 2D $MoS_2$ transistors to reach pressure sensitivity amplification to $\sim 10^3$–$10^7$ $kPa^{-1}$ at an optimized pressure regime of $\sim 1.5$ kPa [29].

Due to the atomic thickness, the electrical properties of two-dimensional materials are highly affected by ambient gases and their pressure variations. The adsorbed gas modifies the electron states within 2D materials changing their electrical conductivity. Owing to the low adsorption energy the process can be reversible.

Specifically, it has been demonstrated that $MoS_2$ conductivity can be enhanced or suppressed by gases such as $O_2$, $CH_4$, $NO_2$, $NO$, $NH_3$, $H_2S$, etc. [22, 30, 31]. Therefore, few- and single-layer $MoS_2$ nanosheets have been investigated for gas and pressure sensing in devices with fast response speed, low power consumption, low minimum pressure detection limits and excellent stability. For instance, few-layer $MoS_2$ back-gate field effect transistors, fabricated on $SiO_2$/Si substrate with Au electrodes, have been demonstrated as resistor-based $O_2$ sensors with sensing performance controllable by the back-gate voltage. Remarkably, these devices have been applied to determine $O_2$ partial pressure with a detectability as low as $6.7 \times 10^{-7}$ millibars at a constant vacuum pressure and proposed as a vacuum gauge [32].

In this paper, we fabricate $MoS_2$ back-gate field effect transistors using $MoS_2$ nanosheets grown by chemical vapor deposition (CVD) on $SiO_2$/Si substrate and measure their electrical characteristics at different air pressures. We show that the threshold voltage of the transistors increases with the increasing pressure. We ascribe such a feature to pressure-dependent adsorption of electronegative oxygen, nitrogen and water molecules, which decrease the n-doping of the MoS$_2$ channel and hence increase the threshold voltage of the transistors. We propose to exploit the dependence of the transistor current on the air pressure to realize vacuum gauges with wide dynamic range and low power consumption.

## 2  Experimental

The MoS$_2$ monolayer flakes were grown by CVD on a heavily doped Si substrate covered by 285 nm SiO$_2$, spin coated with 1% sodium cholate solution. The molybdenum needed for the growth was provided by a saturated ammonium heptamolybdate (AHM) solution, which was annealed at 300 °C under ambient conditions to turn AHM into MoO$_3$. The substrate and the AHM solution were placed in a three-zone tube furnace, along with 50 mg of S powder, positioned upstream in a separate heating zone. The zones containing the S and the AHM were heated to 150 °C and 750 °C, respectively. After 15 min of growth, the process was stopped, and the sample cooled down rapidly.

MoS$_2$ nanoflakes with different shapes and thicknesses, depending on both the local stoichiometry and temperature, were formed [33]. An example is shown in Fig. 1a.

We used optical microscope inspection, with contrast calibrated to approximately estimate the number of layers, to identify MoS$_2$ nanoflakes suitable for the transistor fabrication. A standard e-beam lithography and lift-off process was applied to evaporate Ti (10 nm) and Au (40 nm) bilayers on the flake for the formation of the source and drain electrodes. The back-gate electrode was formed by scratching the Si substrate surface and dropping silver paste.

The SEM top view of a typical device, fabricated using a star-like nanoflake, is shown in Fig. 1a. Figure 1b shows the Raman spectrum of the flake under 532 nm laser excitation. The wavenumber difference, $\Delta k \approx 20.8$ cm$^{-1}$, between the $E_{2g}^1$ (in-plane optical vibration of S atoms in the basal plane) and $A_{1g}$ (out-of-plane optical vibration of S atoms along the c axis) indicates a monolayer [34, 35].

Figure 1c displays the schematic cross-section of the device and the circuits used for the electrical characterization of the transistor in common source configuration. The electrical measurements were carried out inside a cryogenic probe station with fine pressure control (Janis ST 500), connected to a Keithley 4200 SCS (source measurement units, Tektronix Inc.), at room temperature.

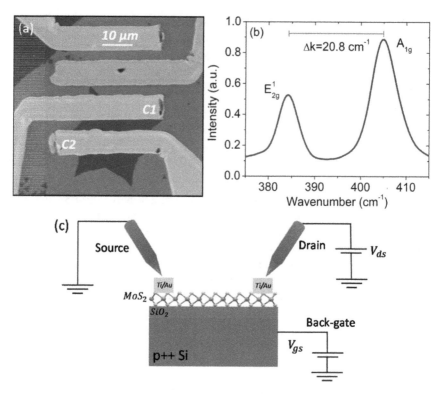

**Fig. 1** **a** SEM image showing the star-like MoS$_2$ nanoflake used as the channel of the back-gate transistor with Ti/Au contacts. The channel width and length are 28.0 μm and 4.4 μm, respectively. **b** Raman spectrum of the MoS$_2$ nanoflake with $E_{2g}^1 - A_{1g}$ wavenumber separation corresponding to a monolayer. **c** MoS$_2$ FET schematic with biasing circuits used for the electrical characterization

## 3 Results and Discussion

Figure 2a, b show the $I_{ds} - V_{ds}$ output characteristics and the $I_{ds} - V_{gs}$ transfer characteristics of the MoS$_2$ transistor measured in high vacuum and at room temperature. As often observed in MoS$_2$ and other 2D-material based devices, the output characteristic exhibits an asymmetric behavior for positive and negative drain biases. As we have demonstrated elsewhere, such a feature is caused by the different contact area as well as by a difference in the Schottky barrier height at the two contacts resulting from local MoS$_2$ processing or intrinsic defects [36, 37]. The transfer characteristic shows a normally-on, n-type transistor. The intrinsic n-type conduction is typical of MoS$_2$ and is mainly due to S vacancies [38]. Compared to similar devices reported in the literature, the transistor shows good metrics in terms of on/off ratio $10^8$ at ±60 V, on-current ∼ 0.3 $\frac{\mu A}{\mu m}$, subthreshold swing of 3.5 $\frac{V}{decade}$ and mobility $\mu = \frac{L}{WC_{ox}V_{ds}} \frac{dI_{ds}}{dV_{gs}} \approx 1.2$ cm$^2$ V$^{-1}$ s$^{-1}$(L and W are the channel length and width,

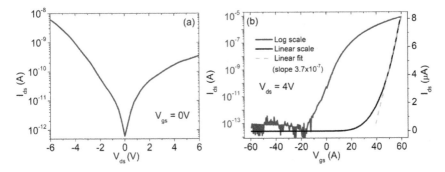

**Fig. 2** **a** Output and **b** transfer (on linear and logarithmic scale) characteristics of the device between C1 and C2 contacts measured at room temperature and $10^{-6}$ Torr pressure. The dashed red line is a linear fit used to evaluate the channel field effect mobility

$V_{ds}$ is the source-drain bias and $C_{ox} = 12.1$ nF cm$^{-2}$ is the SiO$_2$ capacitance per unit area) [39–41].

The result of transfer characteristic measurements at different pressures, P, from high vacuum to atmospheric pressure and back to $10^{-6}$ Torr, is displayed in Fig. 3a. The increasing air pressure causes a right-shift of the transfer curve and therefore an increase of transistor threshold voltage, $V_{th}$. The threshold voltage is here defined as the x-axis intercept of the straight lines fitting the $I_{ds} - V_{ds}$ curves in the current range 1–100 nA. The effect is reversible, in fact the device returns to the pristine state when the high vacuum is restored, as shown by the dash-dot grey line in Fig. 3a. We note that the effect of air pressure on the channel conductance, which could result in the dramatic transformation of n-type to p-type conduction when passing from high vacuum to atmospheric pressure, has been reported also for other 2D TMDs

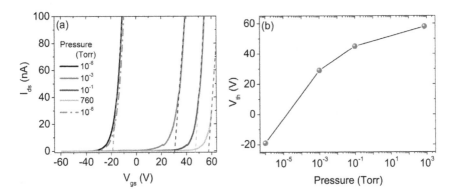

**Fig. 3** **a** Transfer characteristics (solid lines) on linear scale for increasing pressure from high vacuum to atmospheric. The dashed lines are linear fits used to evaluate the transistor threshold voltage. The dash-dot gray line is obtained after that the high vacuum is restored. **b** Threshold voltage as a function of the pressure (color figure online)

materials such as $WSe_2$ or $PdSe_2$ [11, 42]. The effect is usually reversible although it has been found that an aging can occur in specific TMDs, such as $PdSe_2$, after a long (>20 days) air exposure at atmospheric pressure [43].

The monotonic $V_{th} - P$ behaviour, shown in Fig. 3b, suggests that the transistor can be used as pressure sensor, with maximum sensitivity up to $\frac{dV}{d(\log_{10} P)} \approx 13 \frac{V}{decade}$ at lower pressures, where the $V_{th} - P$ curve is steeper. Besides the higher sensitivity, the duty cycle of the device increases when operated in vacuum because of the suppressed air aging effect. Therefore, the sensor is best suited as a vacuum gauge. Moreover, low current of 1 nA or less is needed to monitor the $V_{th}$ variation, which implies that the sensor can be operated in low power-consumption regime.

To investigate the working principle of the device, we measured the transfer characteristics over a gate voltage loop (backward and forward sweep) in air and vacuum. Figure 4a shows the appearance of a hysteresis that diminishes for decreasing pressure. Hysteresis is a well-known phenomenon in transistors with 2D material channels and has been attributed to charge trapping in intrinsic defects of the 2D material, in the gate dielectric and in adsorbate molecules [38, 40, 44–46]. The reducing hysteresis with pressure confirms that adsorbates play an important role in the device under study.

Owing to their high electronegativity, molecular $O_2$, $N_2$ and $H_2O$, adsorbed on $MoS_2$ surface (Fig. 4b), can withdraw electrons from the channel causing the observed increase of the threshold voltage, i.e. of the gate voltage needed to enable conduction in the transistor channel. Absorption occurs particularly at S vacancy sites and the absorption/desorption rate obviously depends on the air pressure the $MoS_2$ nanosheet is exposed to, thus enabling its monitoring.

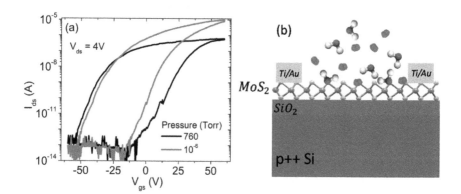

**Fig. 4** **a** Transfer characteristics showing a hysteresis between the forward and reverse $V_{gs}$ sweeps. The hysteresis width decreases for lowering pressure from atmospheric to $10^{-6}$ Torr. **b** Schematic showing the adsorption of molecular $O_2$ and $H_2O$ which, being electronegative, cause the decrease of the electron density in the transistor channel and an increase of the threshold voltage

# 4   Conclusion

We have fabricated and electrically characterized monolayer MoS$_2$ field effect transistors. We have found that the threshold voltage of the transistors increases monotonously with the air pressure, as effect of reduced n-doping caused by adsorption of electronegative O$_2$, N$_2$ and water. Therefore, we have proposed the transistors as air pressure sensors, highlighting their suitability as a vacuum gauge with long duty cycle and low power consumption.

# References

1. Geim AK, Novoselov KS (2007) The rise of graphene. Nat Mater 6:183–191. https://doi.org/10.1038/nmat1849
2. Di Bartolomeo A (2016) Graphene Schottky diodes: an experimental review of the rectifying graphene/semiconductor heterojunction. Phys Rep 606:1–58. https://doi.org/10.1016/j.physrep.2015.10.003
3. Allen MJ, Tung VC, Kaner RB (2010) Honeycomb carbon: a review of graphene. Chem Rev 110:132–145. https://doi.org/10.1021/cr900070d
4. Urban F, Lupina G, Grillo A, Martucciello N, Di Bartolomeo A (2020) Contact resistance and mobility in back-gate graphene transistors. Nano Express 1:010001. https://doi.org/10.1088/2632-959X/ab7055
5. Di Bartolomeo A (2020) Emerging 2D materials and their Van Der Waals heterostructures. Nanomaterials 10:579. https://doi.org/10.3390/nano10030579
6. Novoselov KS, Mishchenko A, Carvalho A, Castro Neto AH (2016) 2D materials and van der Waals heterostructures. Science 353:aac9439. https://doi.org/10.1126/science.aac9439
7. Choi W, Choudhary N, Han GH, Park J, Akinwande D, Lee YH (2017) Recent development of two-dimensional transition metal dichalcogenides and their applications. Mater Today 20:116–130. https://doi.org/10.1016/j.mattod.2016.10.002
8. Lv R, Robinson JA, Schaak RE, Sun D, Sun Y, Mallouk TE, Terrones M (2015) Transition metal dichalcogenides and beyond: synthesis, properties, and applications of single- and few-layer nanosheets. Acc Chem Res 48:56–64. https://doi.org/10.1021/ar5002846
9. Di Bartolomeo A, Urban F, Passacantando M, McEvoy N, Peters L, Iemmo L, Luongo G, Romeo F, Giubileo F (2019) A WSe$_2$ vertical field emission transistor. Nanoscale 11:1538–1548. https://doi.org/10.1039/C8NR09068H
10. Ravindra NM, Tang W, Rassay S (2019) Transition metal dichalcogenides properties and applications. In: Pech-Canul MI, Ravindra NM (eds) Semiconductors. Springer International Publishing, Cham, pp 333–396
11. Urban F, Martucciello N, Peters L, McEvoy N, Di Bartolomeo A (2018) Environmental effects on the electrical characteristics of back-gated WSe$_2$ field-effect transistors. Nanomaterials 8:901. https://doi.org/10.3390/nano8110901
12. Liu K, Yan Q, Chen M, Fan W, Sun Y, Suh J, Fu D, Lee S, Zhou J, Tongay S, Ji J, Neaton JB, Wu J (2014) Elastic properties of chemical-vapor-deposited monolayer MoS$_2$, WS$_2$, and their bilayer heterostructures. Nano Lett 14:5097–5103. https://doi.org/10.1021/nl501793a
13. Cai Y, Xu K, Zhu W (2018) Synthesis of transition metal dichalcogenides and their heterostructures. Mater Res Express 5:095904. https://doi.org/10.1088/2053-1591/aad950
14. Krishnan U, Kaur M, Singh K, Kumar M, Kumar A (2019) A synoptic review of MoS$_2$: synthesis to applications. Superlattices Microstruct 128:274–297. https://doi.org/10.1016/j.spmi.2019.02.005

15. Li X, Zhu H (2015) Two-dimensional $MoS_2$: properties, preparation, and applications. J Materiomics 1:33–44. https://doi.org/10.1016/j.jmat.2015.03.003
16. Iemmo L, Urban F, Giubileo F, Passacantando M, Di Bartolomeo A (2020) Nanotip contacts for electric transport and field emission characterization of ultrathin MoS2 flakes. Nanomaterials 10:106. https://doi.org/10.3390/nano10010106
17. Kwon H, Garg S, Park JH, Jeong Y, Yu S, Kim SM, Kung P, Im S (2019) Monolayer $MoS_2$ field-effect transistors patterned by photolithography for active matrix pixels in organic light-emitting diodes. npj 2D Mater Appl 3:9. https://doi.org/10.1038/s41699-019-0091-9
18. Giubileo F, Iemmo L, Passacantando M, Urban F, Luongo G, Sun L, Amato G, Enrico E, Di Bartolomeo A (2019) Effect of electron irradiation on the transport and field emission properties of few-layer $MoS_2$ field-effect transistors. J Phys Chem C 123:1454–1461. https://doi.org/10.1021/acs.jpcc.8b09089
19. Gant P, Huang P, Pérez de Lara D, Guo D, Frisenda R, Castellanos-Gomez A (2019) A strain tunable single-layer $MoS_2$ photodetector. Mater Today 27:8–13. https://doi.org/10.1016/j.mattod.2019.04.019
20. Di Bartolomeo A, Genovese L, Foller T, Giubileo F, Luongo G, Croin L, Liang S-J, Ang LK, Schleberger M (2017) Electrical transport and persistent photoconductivity in monolayer $MoS_2$ phototransistors. Nanotechnology 28:214002. https://doi.org/10.1088/1361-6528/aa6d98
21. Madauß L, Zegkinoglou I, Vázquez Muiños H, Choi Y-W, Kunze S, Zhao M-Q, Naylor CH, Ernst P, Pollmann E, Ochedowski O, Lebius H, Benyagoub A, Ban-d'Etat B, Johnson ATC, Djurabekova F, Roldan Cuenya B, Schleberger M (2018) Highly active single-layer $MoS_2$ catalysts synthesized by swift heavy ion irradiation. Nanoscale 10:22908–22916. https://doi.org/10.1039/C8NR04696D
22. Li W, Zhang Y, Long X, Cao J, Xin X, Guan X, Peng J, Zheng X (2019) Gas sensors based on mechanically exfoliated $MoS_2$ nanosheets for room-temperature $NO_2$ detection. Sensors 19:2123. https://doi.org/10.3390/s19092123
23. Park M, Park YJ, Chen X, Park Y-K, Kim M-S, Ahn J-H (2016) $MoS_2$-based tactile sensor for electronic skin applications. Adv Mater 28:2556–2562. https://doi.org/10.1002/adma.201505124
24. Zhang W, Zhang P, Su Z, Wei G (2015) Synthesis and sensor applications of $MoS_2$-based nanocomposites. Nanoscale 7:18364–18378. https://doi.org/10.1039/C5NR06121K
25. Kalantar-zadeh K, Ou JZ (2016) Biosensors based on two-dimensional $MoS_2$. ACS Sens 1:5–16. https://doi.org/10.1021/acssensors.5b00142
26. Kim SJ, Mondal S, Min BK, Choi C-G (2018) Highly sensitive and flexible strain-pressure sensors with cracked paddy-shaped $MoS_2$/graphene foam/ecoflex hybrid nanostructures. ACS Appl Mater Interfaces 10:36377–36384. https://doi.org/10.1021/acsami.8b11233
27. Akhter MJ, Kuś W, Mrozek A, Burczyński T (2020) Mechanical properties of monolayer $MoS_2$ with randomly distributed defects. Materials 13:1307. https://doi.org/10.3390/ma13061307
28. Yu F, Liu Q, Gan X, Hu M, Zhang T, Li C, Kang F, Terrones M, Lv R (2017) Ultrasensitive pressure detection of few-layer $MoS_2$. Adv Mater 29:1603266. https://doi.org/10.1002/adma.201603266
29. Huang Y-C, Liu Y, Ma C, Cheng H-C, He Q, Wu H, Wang C, Lin C-Y, Huang Y, Duan X (2020) Sensitive pressure sensors based on conductive microstructured air-gap gates and two-dimensional semiconductor transistors. Nat Electron 3:59–69. https://doi.org/10.1038/s41928-019-0356-5
30. Park J, Mun J, Shin J-S, Kang S-W (2018) Highly sensitive two-dimensional $MoS_2$ gas sensor decorated with Pt nanoparticles. R Soc Open Sci 5:181462. https://doi.org/10.1098/rsos.181462
31. Urban F, Giubileo F, Grillo A, Iemmo L, Luongo G, Passacantando M, Foller T, Madauß L, Pollmann E, Geller MP, Oing D, Schleberger M, Di Bartolomeo A (2019) Gas dependent hysteresis in $MoS_2$ field effect transistors. 2D Mater 6:045049. https://doi.org/10.1088/2053-1583/ab4020
32. Tong Y, Lin Z, Thong JTL, Chan DSH, Zhu C (2015) $MoS_2$ oxygen sensor with gate voltage stress induced performance enhancement. Appl Phys Lett 107:123105. https://doi.org/10.1063/1.4931494

33. Wang L, Chen F, Ji X (2017) Shape consistency of MoS$_2$ flakes grown using chemical vapor deposition. Appl Phys Express 10:065201. https://doi.org/10.7567/APEX.10.065201
34. Gołasa K, Grzeszczyk M, Korona KP, Bożek R, Binder J, Szczytko J, Wysmołek A, Babiński A (2013) Optical properties of molybdenum disulfide (MoS_2). Acta Phys Pol A 124:849–851. https://doi.org/10.12693/APhysPolA.124.849
35. Zeng H, Zhu B, Liu K, Fan J, Cui X, Zhang QM (2012) Low-frequency Raman modes and electronic excitations in atomically thin MoS$_2$ films. Phys Rev B 86:241301. https://doi.org/10.1103/PhysRevB.86.241301
36. Di Bartolomeo A, Grillo A, Urban F, Iemmo L, Giubileo F, Luongo G, Amato G, Croin L, Sun L, Liang S-J, Ang LK (2018) Asymmetric Schottky contacts in bilayer MoS$_2$ field effect transistors. Adv Funct Mater 28:1800657. https://doi.org/10.1002/adfm.201800657
37. Di Bartolomeo A, Giubileo F, Grillo A, Luongo G, Iemmo L, Urban F, Lozzi L, Capista D, Nardone M, Passacantando M (2019) Bias tunable photocurrent in metal-insulator-semiconductor heterostructures with photoresponse enhanced by carbon nanotubes. Nanomaterials 9:1598. https://doi.org/10.3390/nano9111598
38. Di Bartolomeo A, Genovese L, Giubileo F, Iemmo L, Luongo G, Foller T, Schleberger M (2017) Hysteresis in the transfer characteristics of MoS$_2$ transistors. 2D Mater 5:015014. https://doi.org/10.1088/2053-1583/aa91a7
39. Yuan H, Cheng G, You L, Li H, Zhu H, Li W, Kopanski JJ, Obeng YS, Hight Walker AR, Gundlach DJ, Richter CA, Ioannou DE, Li Q (2015) Influence of metal–MoS$_2$ interface on MoS$_2$ transistor performance: comparison of Ag and Ti contacts. ACS Appl Mater Interfaces 7:1180–1187. https://doi.org/10.1021/am506921y
40. Ahn J-H, Parkin WM, Naylor CH, Johnson ATC, Drndić M (2017) Ambient effects on electrical characteristics of CVD-grown monolayer MoS$_2$ field-effect transistors. Sci Rep 7:4075. https://doi.org/10.1038/s41598-017-04350-z
41. Grillo A, Di Bartolomeo A, Urban F, Passacantando M, Caridad JM, Sun J, Camilli L (2020) Observation of 2D conduction in ultrathin germanium arsenide field-effect transistors. ACS Appl Mater Interfaces 12:12998–13004. https://doi.org/10.1021/acsami.0c00348
42. Giubileo F, Grillo A, Iemmo L, Luongo G, Urban F, Passacantando M, Di Bartolomeo A (2020) Environmental effects on transport properties of PdSe$_2$ field effect transistors. Mater Today Proc 20:50–53. https://doi.org/10.1016/j.matpr.2019.08.226
43. Hoffman AN, Gu Y, Liang L, Fowlkes JD, Xiao K, Rack PD (2019) Exploring the air stability of PdSe$_2$ via electrical transport measurements and defect calculations. npj 2D Mater Appl 3:50. https://doi.org/10.1038/s41699-019-0132-4
44. Di Bartolomeo A, Pelella A, Liu X, Miao F, Passacantando M, Giubileo F, Grillo A, Iemmo L, Urban F, Liang S (2019) Pressure-tunable ambipolar conduction and hysteresis in thin palladium diselenide field effect transistors. Adv Funct Mater 29:1902483. https://doi.org/10.1002/adfm.201902483
45. Kaushik N, Mackenzie DMA, Thakar K, Goyal N, Mukherjee B, Boggild P, Petersen DH, Lodha S (2017) Reversible hysteresis inversion in MoS$_2$ field effect transistors. npj 2D Mater Appl 1:34. https://doi.org/10.1038/s41699-017-0038-y
46. Bartolomeo AD, Rinzan M, Boyd AK, Yang Y, Guadagno L, Giubileo F, Barbara P (2010) Electrical properties and memory effects of field-effect transistors from networks of single- and double-walled carbon nanotubes. Nanotechnology 21:115204. https://doi.org/10.1088/0957-4484/21/11/115204

# Multipurpose Platform for Piezoelectric Materials and Devices Characterization

M. F. Bevilacqua, R. Scaldaferri, and I. Pedaci

**Abstract** Microelectromechanical systems (MEMS) are one of the most peculiar examples of applications of piezoelectric materials. Piezo-materials exhibit characteristics that allow to use them as for both sensing and actuating. MEMS division represents one of the most significant business department of ST. In this context, the activities of AMS R&D team of Naples are focused on piezoelectric characterization frame ranging from material up to final package applications. This work is aimed to provide an overview of different electro-mechanical and reliability characterization strategies implemented to deepen understanding piezo-materials and device's properties.

**Keywords** MEMS · Polarization · Laser Doppler vibrometry · BDV · Weibull · Statistic on wafer · Reliability

## 1 Introduction

MEMS-sensors and actuators, based on PZT (Lead Zirconate Titanate) thin films, have collected significant attention for the large number of related applications. Sensors exploit direct piezoelectric effect transforming mechanical input into electrical output (accelerometers, gyroscope, pressure sensors, microphones) while actuators are based on inverse piezoelectric effect because of which an electrical input is transformed into a mechanical output (micromirrors, piezo actuators) [1].

In the field of ST piezo-based product portfolio we have focused on a wide characterization chain ranging from material's stack up to final applications.

Within this framework, ferroelectric, dielectric and piezoelectric properties of thin films are investigated to provide an insight into material characteristics, interfaces and dynamic behavior. A deep comprehension of these aspects allows to select main parameters affecting device operation and discard the irrelevant ones. As integration to this process of piezo-material quality, reliability margins and defectivity screening

M. F. Bevilacqua (✉) · R. Scaldaferri · I. Pedaci
STMicroelectronics, Viale Remo de Feo 1, 80022 Arzano, Naples, Italy
e-mail: maria-fortuna.bevilacqua@st.com

© The Author(s), under exclusive license to Springer Nature Switzerland AG 2021
G. Di Francia and C. Di Natale (eds.), *Sensors and Microsystems*,
Lecture Notes in Electrical Engineering 753,
https://doi.org/10.1007/978-3-030-69551-4_8

of final devices are investigated. By this way the prevision of device's behavior under specific stresses such as temperature, bias or, eventually, gases exposures are carried out by using dedicated equipment.

With this paper we wish to give you a panorama of a part of measurements that we can perform and the philosophy of this integrated approach.

## 2 Experimental and Results

### 2.1 *Experimental Setup to Study Material Properties*

The characterization of piezoelectric material and electrodes interfaces goes through the analysis of the electrical and dielectric properties of material and devices [2]. With this aim leakage current at fixed voltage (CVS), even at different temperatures, can be a fundamental tool to give information about insulating properties of piezoelectric material and interfaces. The test above mentioned was done by MPI SUMMIT 200 Semiautomatic probe station, Keysight B1500A Semiconductor Device Parameter Analyzer and 6487A Keithley picoammeter (Fig. 1a).

To explore the capability to transduce an electrical input into a mechanical actuation, i.e. the piezo efficiency of material's stack, we perform a Laser Doppler Vibrometry analysis with a Polytec MSA 500 vibrometer (Fig. 1b). In this way we can extract the maximum actuation, and, through FFT measurements, the modal analysis from frequency spectrum.

Ferroelectricity of piezo-based materials is tested by Aixact Thin Film Analyzer, TFA2000 (Fig. 2). This equipment gives out information about ferroelectric

a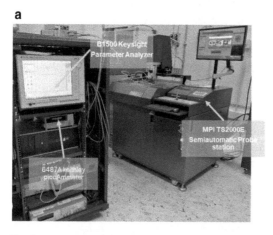

b

**Fig. 1** **a** Pictures of MPI semiautomated probe station and picoammeter. **b** Polytec MSA 500 laser Doppler vibrometer

**Fig. 2** Aixact thin film
analyzer (TFA2000) for
ferroelectric analysis

hysteresis loop of piezo layer. As particular application of ferroelectric analysis
we report the Dielectric Hysteresis Measurement (DHM), i.e. the polarization loop,
that provides info about the switching of ferroelectric domains under application of
an electric field [3, 4].

All measurements described in this paper have been carried out on devices made
of a multi-materials stack including a PZT layer of 2 µm deposited between top and
bottom electrodes.

## 2.2  Experimental Setup and Statistic on Wafer

Through the measurement of I–V curves at different temperatures it is possible
to detect the maximum voltage that DUT (Device Under Test) can suffer, named
Breakdown Voltage (BV). The analysis of BV Distributions (BVD) on wafer allows
to identify the nature of failure mechanisms. BVD analysis is performed by using
the same setup described in Sect. 2.1 and showed in Fig. 1a.

By this analysis, intrinsic failures related to the process and to the materials can
be identified and studied and, in the meantime, extrinsic failure mechanisms could
be recognized.

## 2.3 Reliability and Experimental Setup

Reliability tests concern the ability of our components to perform a given function under stated conditions for the desired duration. To explore these topics, we can perform different reliability tests to accelerate the lifetime of MEMS by using acceleration factors for proper lifetime prediction.

The accelerated lifetime test is a method to estimate the failure rate of a component during actual use. It focuses on the specific stresses suffered by the component in the environment in which it is used. This test is performed by using stress conditions as parameters (voltage, temperature, humidity) and it is widely used when developing new products that adopted a new process. This analysis was carried out through static test such as Time Dependent Dielectric Breakdown (TDDB).

TDDB is an important failure mechanism where dielectric failures occur as soon as a conductive path shorten the electrodes. Under the influence of an external electric field, a leakage current exists. If the electrons are energetic enough to damage the dielectric material, they can create additional defect states. To determine the time to breakdown, the leakage current under a fixed electrostatic potential stress is measured as a function of time. The time taken to reach this point is the dielectric time-to-breakdown, $\tau_{bd}$, and Weibull statistics are then used to analyze the failure data and determine the characteristic lifetimes [5].

In general, the lifetimes are predicted from the extrapolation of experimental results, which are measured under accelerated conditions (i.e. at temperature and stress voltage higher than actual operating conditions). TDDB on piezo-capacitors is achieved by our Qualitau-Infinity System, custom equipment for TDDB test (Fig. 3).

TDDB is a statistical phenomenon, therefore, it can only be described in statistical terms. Weibull distribution is the most suitable for the description of $\tau_{bd}$.

$$F = 1 - \exp\left[-\left(\frac{\tau_{bd}}{\eta}\right)^{\beta}\right]$$

where $F$ is the cumulative failure probability, $\eta$ and $\beta$ represent, respectively, the scale and shape parameters. $\eta$ is related to the characteristic lifetime while the shape factor $\beta$ is related to the physical defects leading to breakdown [6]. The temperature is a factor that strongly affects the $\tau_{bd}$ of the devices. This analysis gives a feedback about failure mechanisms and allows to forecast lifetime of desired structures.

## 2.4 Results

With a view to the analysis of material's properties a typical CVS is shown in Fig. 4a where the current at fixed voltage is recorded as function of time. This measure allows to recognize typical decay time and to observe resistance degradation effect typical of PZT.

**Fig. 3** Qualitau-infinity
system for TDDB reliability
tests

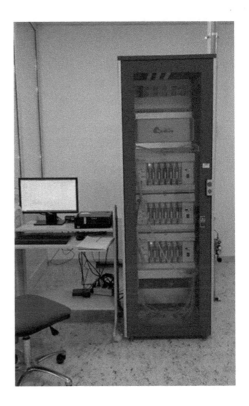

Moreover, from the point of view of material characterization, the P–V measure allows to estimate remnant polarization and coercive field that are determining factors to evaluate the goodness of piezoelectric performances. In Fig. 4b, a typical P–V curve is reported by applying to the device a large bipolar triangular signal, in the range 0–30 V at the frequency of 100 Hz.

Figure 4c shows the FFT plot of a moving piezo structure acquired by applying a periodic chirp signal with 1 V amplitude and 0 V offset and collecting the resulting frequency spectrum in the range from 1 to 40 kHz.

From experimental data in Fig. 4a it is possible to observe that CVS curve exhibits the characteristic trend of piezo-materials where at 180 °C and 90 V on 2 $\mu$m PZT the leakage current is firstly due to the contribution of resistance degradation and restoration and then is mainly dominated by relaxation phenomena [7].

Polarization curve shown in Fig. 4b has a saturation value of 22 $\mu$C/cm$^2$ and an asymmetric loop with a shift along-E axis due to the existence of an internal field. This is mostly due to the combination of a thermal and electrical treatment, "poling" of piezo, to which the material underwent to.

The FFT plot in Fig. 4c allows to determine all the vibrating modes (harmonics) related to the different motion directions of the device. This parameter is associated to the elastic properties of piezo-material and to the device design.

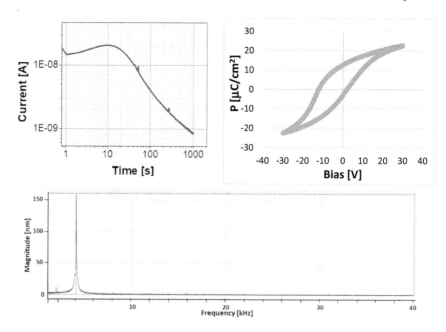

**Fig. 4** **a** Plot of leakage current versus time of 2 μm thick PZT, collected at 180 °C with a DC bias of 90 V. **b** Dielectric hysteresis loop versus applied bias shows the bipolar behavior of polarization. **c** FFT spectrum of a free moving piezo membrane where the first harmonic is visible

As specified in the previous paragraph, our test platform consists also of the analysis of BDV distribution on wafers. In Fig. 5a, a BDV distribution of DUTs selected from a single wafer is collected for different polarization directions (positive

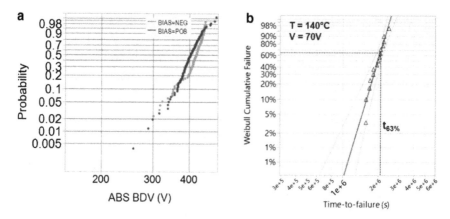

**Fig. 5** **a** BDV distribution of samples for different tests with positive and negative bias (blue and red circles respectively). **b** Weibull cumulative failure distribution for a test performed at 140 °C and 70 V. (color figure online)

indicated by blue and negative by red circles). The measure is performed by applying a voltage ramp from 0 V to a maximum voltage and detecting the BDV as the voltage corresponding to a current exceeding a fixed value defined as "failure criterion". When this condition occurs device burning, cracks and electrodes breakage could happen. The distribution of the occurrences of these failures on the wafer could be indicative of failure mechanisms and give a feedback on processes and materials.

In Fig. 5b we show a Weibull Cumulative Failure distribution for a population of DUTs tested at 140 °C and 70 V under CVS stress. The characteristic lifetime, $\eta$, is the time-to-breakdown that corresponds to 63.2% of cumulative failure percentage, i.e.:

$$F(\tau_{bd}) = 63.2\%$$

This distribution shows a good $\beta$ value due to data points well aligned to the fit line. This behavior is typical of a single degradation mechanism that leads to device failure. By changing test conditions, bias and temperature, the values of $\tau_{bd}$ and $\beta$ change giving out the precious information that time to failure decreases with increasing electric field or temperature.

## 3    Conclusion

With this paper we meant to give an overview of a part of all possible investigations on piezo materials, statistical analysis and reliability tests that we can perform in our ST labs, with the aim to deeply understand piezo conduction models and failure mechanisms occurring on a large number of samples. The analysis of leakage current behavior allowed to underline resistance degradation and restoration effects together with the exploration of material-interface properties. The broken symmetry of PV loop let us explore and assess the effect of thermo-electric treatments. In this way, stable poling conditions, which produce expected effects on mechanical and ferroelectric parameters, can be determined.

In the section of statistical analysis, BDV distribution on wafer was investigated. This represents an important screening procedure to identify process weaknesses. At the same time the analysis of dielectric properties on wafer allows to investigate the interfaces properties, the effects of passivation layers and the insulating properties of piezo-materials.

Reliability tests were designed to reproduce failure mechanisms under accelerated conditions. This is extremely important to know the product reliability in its real use. The effect of stresses such as temperature or voltage have been demonstrated to play a key role in the occurrence and understanding of failure mechanisms.

# References

1. AME Microsystems Product Marketing (2017) MEMS micro-actuators enabling new and unforeseen applications. STMicroelectronics
2. Damjanovic D, Mayergoyz I, Bertotti G (2005) Hysteresis in piezoelectric and ferroelectric materials. In: The science of hysteresis, vol 3. Elsevier, Amsterdam
3. Koval V, Viola G, Tan Y (2015) Biasing effects in ferroic materials. In: Ferroelectric materials—synthesis and characterization. Intech, pp 205–247
4. Rhun G, Bouregba R, Poullain G (2004) Polarization loop deformations of an oxygen deficient Pb(Zr0.25, Ti0.75)O$_3$ ferroelectric thin film. J Appl Phys 96:5712–5721
5. (2017) Semiconductor reliability handbook. Renesas Electronics
6. Qi Y, Zhu Y, Zhang J, Lin X, Cheng K, Jiang L, Yu H (2018) Evaluation of LPCVD SiNx gate dielectric reliability by TDDB measurement in Si-substrate-based AlGaN/GaN MIS-HEM. IEEE Trans Electron Devices 65:1759–1764
7. Defay E (2011) Integration on ferroelectric and piezoelectric thin films. Wiley, Hoboken

# Magnetic Field Detection by an SPR Plastic Optical Fiber Sensor and Ferrofluids

**Nunzio Cennamo, Francesco Arcadio, Luigi Zeni, Aldo Minardo, Bruno Andò, Salvatore Baglio, and Vincenzo Marletta**

**Abstract** Several surface plasmon resonance (SPR) sensors in optical fibers have been presented for the detection of magnetic fields. These SPR sensors measure the variation of the refractive index of a magnetic fluid in contact to the metal film, when the magnetic field changes. In this work, we have presented a different approach to monitor a magnetic field exploiting SPR optical fiber sensors and ferrofluids. More specifically, we have inserted a patch of multimode plastic optical fiber (POF) covered with ferrofluids between a light source and an SPR-POF sensor connected to a spectrometer. The magnetic field in the region near the patch covered by ferrofluids exerts a bending force on the patch changing the light in input to the SPR sensor and modifying the SPR phenomenon, in terms of resonance's shape and wavelength.

**Keywords** Magnetic field sensors · Plastic optical fibers · Surface plasmon resonance · Magnetic fluid · Multimode waveguides

## 1 Introduction

Fiber optic sensors have gained increasing interest from both the scientific and industrial communities for their features like the reduced dimensions, the robustness to harsh environments, no need for electric power supply, the immunity to electromagnetic interference and the possibility of implementing remote sensing.

Recently, the advantages of fiber optic technology have been exploited in the development of magnetic field sensors. The measurement of the magnetic field is a fundamental task in many applications as, for example, current measurements [1–3], biomedical detection [4], localization [5], geophysical research [6], safety and

N. Cennamo (✉) · F. Arcadio · L. Zeni · A. Minardo
Department of Engineering, University of Campania "Luigi Vanvitelli", 81031 Aversa, Italy
e-mail: nunzio.cennamo@unicampania.it

B. Andò · S. Baglio · V. Marletta
Department of Electrical Electronic and Information Engineering (DIEEI), University of Catania, Catania, Italy
e-mail: bruno.ando@unict.it

© The Author(s), under exclusive license to Springer Nature Switzerland AG 2021
G. Di Francia and C. Di Natale (eds.), *Sensors and Microsystems*,
Lecture Notes in Electrical Engineering 753,
https://doi.org/10.1007/978-3-030-69551-4_9

surveillance applications [7]. However, most of the available sensors are sensitive to the presence of magnetic noise [8].

Different solutions exploiting magnetic fluids combined with optical fiber gratings, interferometry, Surface Plasmon Resonance (SPR), and other solutions involving tailored fibers (etched, tapered and U-shaped), have been reported in the scientific literature [9]. In particular, the SPR phenomenon has been proven advantageous in a large range of applications, such as in the selective sensing of substances. The remote sensing capabilities offered by optical fibers could be exploited for on-site, low-cost, and real-time monitoring of different analytes when receptors are present at the metal-dielectric interface [10–17].

Recently, SPR optical fibers have been adopted as magnetic field sensors [9, 18–20]. In particular, the proposed solutions adopt SPR optical fiber sensors to monitor the refractive index of magnetic fluid (ferrofluid) which changes with the target magnetic field. Moreover, low-cost SPR D-shaped Plastic Optical Fiber (POF) sensors emerged for their advantages and flexibility compared to silica optical fibers [21–25].

The magnetic fluids are stable colloidal suspensions constituted by magnetic nanoparticles (such as $Fe_3O_4$) dispersed in an organic solvent (hydrocarbon oils) or water [9] and coated with surfactant agents to prevent their aggregation. Magnetic fluids possess many interesting magneto-optic properties that can be used for sensing [26–28]. Several studies about these magneto-optic properties are reported as, for example, tunable transmittance, birefringence, tunable refractive index, the Faraday effect, and many others [9, 29–34].

In this work, a multimode SPR D-shaped POF coupled with a magnetic fluid to measure static magnetic fields is proposed. Unlike other solutions available in the literature, where the SPR-POF is used to monitor the refractive index of the magnetic fluid under the effect of the target magnetic field [18–20], in the proposed solution the ferrofluid covers a patch of POF and the magnetic field interacts with the POF generating a bending and consequently a change in the resonant conditions of the SPR POF. This working principle represents the main novelty of this work.

The main advantages of the proposed solution are inherited from the use of the optical fibers and the high sensitivity of the SPR phenomenon.

## 2   The Developed Prototype

The sensor prototype consists of an SPR-POF platform, a POF (patch) covered by the ferrofluid and a light source. An outline of the magnetic field Sensor System is shown in Fig. 1.

The SPR-POF sensor consists of a Poly(methyl methacrylate) (PMMA) core of 980 μm and on a fluorinated polymer cladding of 10 μm. The process adopted to obtain the D-shape by removing the cladding has been described in [21]. The D-shaped area has a length of 1 cm. A photoresist S1813 was spun on the sensing area

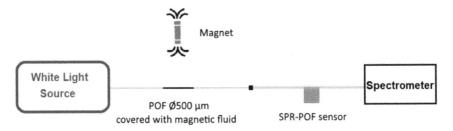

**Fig. 1** Outline of the magnetic field sensor system

and covered by a 60 nm thick gold film sputtered by using a Bal-Tec SCD 500 sputtering machine. The photoresist overlayer permits to improve the SPR phenomenon [21].

The SPR-POF sensor is connected to a POF patch based on a PMMA core of about 480 μm and a fluorinated polymer cladding of 10 μm (total diameter of 500 μm), covered by ferrofluid for about 2 cm in length (see Fig. 1). The patch is used to connect the SPR sensor to a white light source HL–2000–LL by Ocean Optics, with a spectral emission of the lamp in the range from 360 to 1700 nm, and to modify the light at the input of the SPR-POF sensor by varying the distance of a magnet from the coated POF.

The ferrofluid adopted in this work is the MFR-DP1 by MAGRON co., LTD [35].

To monitor the sensor's output, a spectrometer FLAME-S-VIS-NIR-ES (Ocean Optics), directly connected to a computer, was adopted. Measurements were performed in the range from 350 to 1000 nm.

## 3  Experimental Results

In order to test the magnetic field Sensor System we have evaluated how the resonance wavelength changes by keeping fixed the refractive index of the solution upon the SPR-POF sensor and by varying the distance between the magnet and the POF patch covered with ferrofluid.

Figure 2 shows the SPR spectra obtained at a fixed refractive index (1.332) for the configuration without magnet and with magnet at different distances (ranging from 6 to 4 cm) from the POF covered with the magnetic fluid.

Particularly, at the approach of the magnet to the POF patch the resonance wavelength shifts to higher values.

Figure 3 reports the resonance wavelength as a function of the magnetic field, along with the linear fitting to the experimental data. Each experimental value is the average of 5 subsequent measurements and the respective standard deviations (error bars), are shown too. The magnetic field values corresponding to the distances have been estimated with a commercial Hall effect sensor (SS496A, Honeywell).

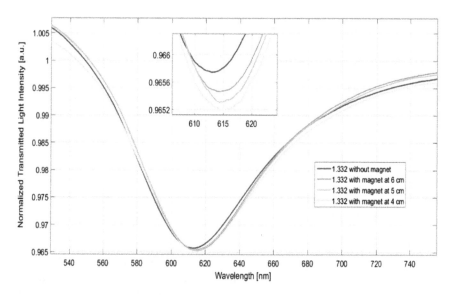

**Fig. 2** SPR normalized transmitted spectra for configuration without magnet and with magnet at different distances (from 6 to 4 cm)

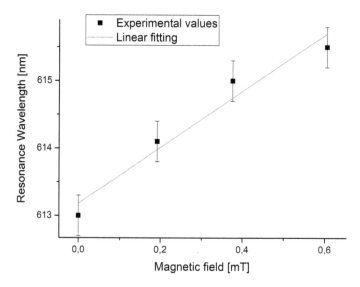

**Fig. 3** Resonance wavelength versus magnetic field and linear fitting to the experimental values

The performances of this kind of sensors are usually determined by parameters like sensitivity and resolution [21]. In such a case, if an alteration of the magnetic field ($\delta M$) produces a variation of the resonance wavelength ($\delta\lambda$), sensitivity at a fixed refractive index (n) can be defined as:

$$S(n) = \frac{\delta \lambda}{\delta M} \text{ [nm/mT]} \qquad (1)$$

From Eq. 1 and Fig. 3 appears that the sensitivity is about 4 nm/mT = 4000 pm/mT. It's important to underline that the sensor's response is strongly dependent on the range of distances. In this case we have chosen a distance ranging from 6 to 4 cm.

Finally the resolution, calculated considering the noise level at the output of the sensor in the configuration without magnet (about 0.3 nm), is equal to about 50 µT.

## 4 Conclusions

In this work we have shown how it is possible to use an SPR-POF sensor as magnetic field sensor. In particular, we have proposed a proof of concept of a novel magnetic field sensing method based on the optical characteristics of multimode POFs.

The obtained results have demonstrated the feasibility of this method, showing very good sensitivity and resolution.

## References

1. Andò B et al (2019) A fluxgate-based approach for ion beam current measurement in ECRIS beamline: design and preliminary investigations. IEEE Trans Instrum Meas 68(5):1477–1484
2. Musuroi C et al (2020) High sensitivity differential giant magnetoresistance (GMR) based sensor for non-contacting DC/AC current measurement. Sensors 20:323. https://doi.org/10.3390/s20010323
3. Andò B et al (2019) Polymeric transducers: an inkjet printed B-field sensor with resistive readout strategy. Sensors 19(23), Article number 5318
4. Shi D et al (2015) Photo-fluorescent and magnetic properties of iron oxide nanoparticles for biomedical applications. Nanoscale 7:8209–8232
5. Gao X et al (2017) Localization of ferromagnetic target with three magnetic sensors in the movement considering angular rotation. Sensors 17:2079
6. Poliakov SV et al (2017) The range of induction-coil magnetic field sensors for geophysical explorations. Seismic Instrum 53:1
7. Wahlström N, Hostettler R, Gustafsson F, Birk W (2014) Classification of driving direction in traffic surveillance using magnetometers. IEEE Trans Intell Transp Syst 15(4):1405–1418
8. Andò B et al (2005) Residence times difference fluxgate. Measurement 38(2):89–112
9. Alberto N et al (2018) Optical fiber magnetic field sensors based on magnetic fluid: a review. Sensors 18:4325
10. Cennamo N et al (2019) A novel sensing methodology to detect furfural in water, exploiting MIPs, and inkjet-printed optical waveguides. IEEE Trans Instrum Meas 68(5):1582–1589
11. Gupta BD, Verma RK (2009) Surface plasmon resonance-based fiber optic sensors: principle, probe designs, and some applications. J Sens 2009:1–12
12. Wang XD, Wolfbeis OS (2013) Fiber-optic chemical sensors and biosensors (2008–2012). Anal Chem 85:487
13. Wang XD, Wolfbeis OS (2016) Fiber-optic chemical sensors and biosensors (2013–2015). Anal Chem 88:203–227

14. Trouillet A et al (1996) Chemical sensing by surface plasmon resonance in a multimode optical fibre. Pure Appl Opt 5:227
15. Jin Y, Granville AM (2016) Polymer fiber optic sensors—a mini review of their synthesis and applications. J Biosens Bioelectron 7:1–11
16. Leung A et al (2007) A review of fiber-optic biosensors. Sens Actuators B Chem 125:688–703
17. Anuj K et al (2007) Fiber-optic sensors based on surface Plasmon resonance: a comprehensive review. IEEE Sens J 7:1118
18. Zhou X et al (2019) Magnetic field sensing based on SPR optical fiber sensor interacting with magnetic fluid. IEEE Trans Instrum Meas 68:234–239
19. Rodríguez-Schwendtner E et al (2017) Plasmonic sensor based on tapered optical fibers and magnetic fluids for measuring magnetic fields. Sens Actuators A Phys 264:58–62
20. Liu H et al (2018) Temperature-compensated magnetic field sensor based on surface plasmon resonance and directional resonance coupling in a D-shaped photonic crystal fiber. Optik 158:1402–1409
21. Cennamo N et al (2011) Low cost sensors based on SPR in a plastic optical fiber for biosensor implementation. Sensors 11:11752
22. Cennamo N et al (2014) High selectivity and sensitivity sensor based on MIP and SPR in tapered plastic optical fibers for the detection of L-nicotine. Sens Actuators B Chem 191:529
23. Cennamo N et al (2013) An innovative plastic optical fiber-based biosensor for new bio/applications. The case of celiac disease. Sens Actuators B Chem 176:1008
24. Cennamo N et al (2014) A simple small size and low cost sensor based on surface plasmon resonance for selective detection of Fe(III). Sensors 14:4657–4671
25. Cennamo N et al (2016) Markers detection in transformer oil by plasmonic chemical sensor system based on POF and MIPs. IEEE Sens J 16:7663–7670
26. Andò B et al (2013) Ferrofluids and their use in sensors. In: Iniewski K (ed) Smart sensors for industrial applications. CRC Press, Boca Raton, pp 355–368
27. Musumeci RE et al (2015) Measurement of wave near-bed velocity and bottom shear stress by ferrofluids. IEEE Trans Instrum Meas 64(5):1232–1239
28. Musumeci RE et al (2018) A ferrofluid-based sensor to measure bottom shear stresses under currents and waves. J Hydraul Res 56(5):630–647
29. Martinez L et al (2005) A novel magneto-optic ferrofluid material for sensor applications. Sens Actuators A Phys 123–124:438–443
30. Yang S et al (2004) Origin and applications of magnetically tunable refractive index of magnetic fluid films. Appl Phys Lett 84:5204–5206
31. Horng H et al (2003) Designing the refractive indices by using magnetic fluids. Appl Phys Lett 82:2434–3436
32. Chen Y et al (2003) Thermal effect on the field-dependent refractive index of the magnetic fluid film. Appl Phys Lett 82:3481–3483
33. Mailfert A et al (1980) Dielectric behavior of a ferrofluid. IEEE Trans Magn 16:254–257
34. Zhao Y et al (2014) Tunable characteristics and mechanism analysis of the magnetic fluid refractive index with applied magnetic field. IEEE Trans Magn 50:4600205
35. https://www.supermagnete.it/eng/school-magnets/ferrofluid-10-ml_M-FER-10. Available on line, accessed on 15 Apr 2020

# Health Monitoring of Brushless Motors for Unmanned Aircraft Systems Through Infrared Thermography

**Simone Boccardi** ⓘ**, Gennaro Ariante** ⓘ**, Umberto Papa** ⓘ**,
Giuseppe Del Core** ⓘ**, and Carosena Meola** ⓘ

**Abstract** A preliminary investigation is performed in order to verify the exploitability of infrared thermography as monitoring tool for predictive fault analysis of brushless electric motors used in small Unmanned Aircraft Systems (UASs). Laboratory tests have been performed by monitoring with an infrared camera some brushless motors while were working at different conditions. The acquired data have been postprocessed through specific techniques and analyzed in order to extract useful information about the engine thermal behavior.

**Keywords** Brushless motors · Infrared thermography · Non-destructive testing · Unmanned aircraft systems

## 1 Introduction

In the last ten years, the use and the spread of small Unmanned Aircraft Systems (UASs) has grown exponentially. The UASs are employed for both civil and military applications, especially in tasks that for men are DDD (Dull-Dirty-Dangerous) [1].

S. Boccardi (✉) · G. Ariante · U. Papa · G. Del Core
Department of Science and Technology, Università degli Studi di Napoli "Parthenope", Naples, Italy
e-mail: simone.boccardi@uniparthenope.it

G. Ariante
e-mail: gennaro.ariante@uniparthenope.it

U. Papa
e-mail: umberto.papa@uniparthenope.it

G. Del Core
e-mail: giuseppe.delcore@uniparthenope.it

C. Meola
Departement of Industrial Engineering, Università degli Studi di Napoli "Federico II", Naples, Italy
e-mail: carosena.meola@unina.it

© The Author(s), under exclusive license to Springer Nature Switzerland AG 2021    69
G. Di Francia and C. Di Natale (eds.), *Sensors and Microsystems*,
Lecture Notes in Electrical Engineering 753,
https://doi.org/10.1007/978-3-030-69551-4_10

At the same time, the new regulations about environmental safeguard are driving the interest of the academic and industrial communities towards the research and development of new applications of the electrical devices in aerospace, especially for the propulsion of manned and unmanned aerial vehicles of different weight and dimensions. Furthermore, in the near future, the use of electrical motors in aeronautics will not be limited to propulsion systems, but it will be increasingly extended to actuators, today mainly hydraulic [2]. In this context, infrared thermography can be regarded as a useful tool for both predictive fault detection and design purposes of almost all the electromechanical components with particular attention to electrical engines. Infrared thermography is a non-invasive method of inspection, therefore it is currently usefully exploited in a wide range of applications, such as medicine, civil and industrial engineering and many other applications [3]. In the aerospace field, infrared thermography has proved its effectiveness as non-destructive technique, to detect different kinds of production defects in composite materials as well damages occurring during their operating life [4, 5]. However, its potentiality has not been yet fully exploited. Vice versa, in the electrical field infrared thermography is more widely used for predictive maintenance and fault detection of electrical equipment such as electrical panels, high voltage distribution network, microelectronics [3, 6] and induction motors for industrial applications [3, 7]. In the latter components, the presence of high and abnormal temperature values with respect to a reference standard one is in general indicative of potential problems. In order to quantify the problem seriousness from the observed temperature values, the International Electrical Testing Association (NETA) proposed a standard where predefined temperature variations $\Delta T$, with respect to a reference value, correspond to levels of maintenance priority [8] (Table 1).

The aim of this paper is to perform a preliminary investigation in order to verify if infrared thermography standards and procedures, currently applied to monitor electrical equipment and induction motors, are suitable to monitor the performance of a small brushless motor (BL motor) designed for small UAVs.

**Table 1** Standard for infrared inspection of electrical equipment

| Temperature difference ($\Delta T$) for similar components under similar loadings (°C) | Temperature difference ($\Delta T$) between component and ambient temperature (°C) | Recommended maintenance action |
|---|---|---|
| 1–3 | 1–10 | Possible deficiency; warrants investigation |
| 4–15 | 11–20 | Probable deficiency; repair as time permits |
| – | 21–40 | Monitor until corrective measures can be accomplished |
| >15 | >40 | Major discrepancy; immediate maintenance |

## 2 Experimental and Results

### 2.1 Experimental Setup

A small BL motor (Turnigy C2830-1050) is constrained over a support that allows for optical access of the infrared camera. The support is properly conceived and realized with the double aim to prevent displacement of the electromechanical motor when it is power on and to allow optical access to monitor almost all its external components. Besides a cylindrical portion of the external metallic surface of the BL motor has been covered with an opaque white paint to increase its emissivity, $\varepsilon$ [3] during some tests. The BL motor is connected to the electronic speed controller (ESC) powered by a lithium polymer battery (LIPO) and managed through a Raspberry Pi Zero W, used to supply the Pulse Width Modulation (PWM) control signal to the electric engine via software properly developed. The infrared camera Flir T650sc, equipped with a micro bolometric LW (long wave) sensor with a full frame option of 640 × 480 pixels, is placed at the minimum focusing distance and used to monitor the electrical engine while it is operating Fig. 1. Several tests have been performed by changing the electric motor operative conditions and the infrared camera relative position. For each of the experimental tests, the infrared camera acquires images with a sampling rate of 15 or 30 Hz while the motor is turned on, according to the following procedure:

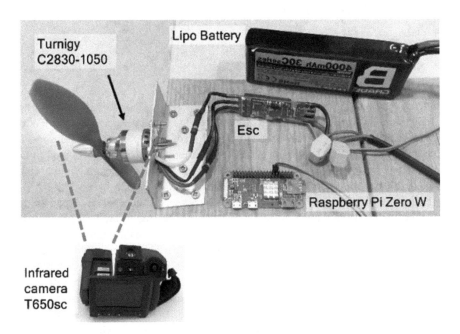

**Fig. 1** Test setup

1   the infrared camera starts to acquire images 8 s before the engine is turned on, in order to have some images at initial environmental conditions;
2   the engine is turned on at the desired PWM value and is left running for about 180 s;
3   the engine is turned off;
4   the infrared camera continues recording images for about 40 s after the engine has been turned off.

## 2.2   Data Analysis

The recorded sequences of images are analyzed to get values of absolute temperature and/or temperature variation $\Delta T$ with respect to the environment one's, in order to qualitatively and quantitatively evaluate the engine thermal behavior over the time while it is working. In particular, to extract data in terms of $\Delta T$ sequences, thermal images are post-processed according to the Eq. (1):

$$\Delta T = T(i, j, t) - T(i, j, 0) \tag{1}$$

where $i$ and $j$ represent lines and columns of the surface temperature array respectively and $t$ the time instant at which one image is recorded; more specifically, $T(i, j, 0)$ indicates the image of the sequence, before starting of the load, for which the engine surface is at ambient temperature.

## 3   Simulations and Results

Some of the obtained results are shown in this section in terms of IR images and $\Delta T$ plots. In Fig. 2 are shown same thermal images for a test performed at PWM 60%. It is clearly possible to underline the variation of the apparent temperature with time on different external components of the BL motor at environmental conditions immediately before the engine is turned on (Fig. 2a), while the engine is working (Fig. 2b–e) and over 180 s when the engine is turned off (Fig. 2f–h).

Looking at Fig. 2 it is clearly possible to recognize the expected heating up effect occurring when the engine starts (Fig. 2b–e) and due to the heat production linked with electro-mechanical dissipative phenomena. When the engine is turned off the apparent temperature continues to rise (Fig. 2f–g). This apparently unexpected evidence can be explained by considering the strong reduction of the convective heat coefficient occurring when the air flow induced by the propeller stops: in the next seconds after the engine is turned off the heat over the external surfaces is no longer efficiently dissipated, and this causes temperature rise. A more quantitative evaluation of the temperature evolution on the BL motor is shown in Fig. 3 where are reported the mean $\Delta T$ value evaluated inside the black rectangle (Fig. 3a), placed on the

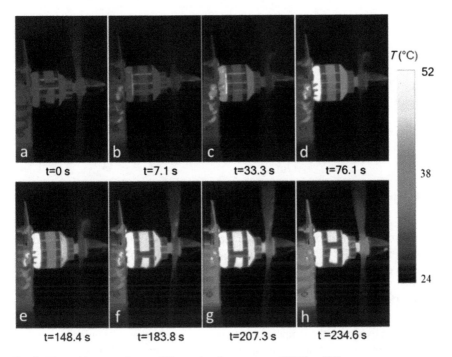

**Fig. 2** Thermal images taken at different time for a constant PWM = 60%

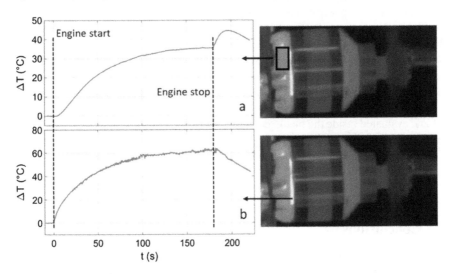

**Fig. 3** $\Delta T$ over time; **a** mean value inside the black rectangle; **b** maximum value over the whole image for a PWM = 60%

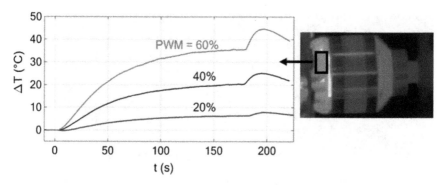

**Fig. 4** Variation of the mean ΔT (inside the black rectangle) for some values of PWM

backside on a non-rotating part of the engine, and the maximum $\Delta T$ over the whole engine (Fig. 3b). For all the performed tests, the maximum values of temperature variation over time were always attained in correspondence of the very small gap (less than 1 mm) between the rear non-rotating part and the front rotating part of the engine. This region corresponds to the vertical yellow line in Fig. 2. In fact, through the small gap the infrared camera directly views the final small portion of the stator winding. As expected, the temperature starts to rise as soon as the engine is switched on and continues to rise until a quasi-stationary constant value is reached. As soon as the engine is stopped the mean $\Delta T$ starts to rise again on the engine external components (Fig. 3a) and then smoothly decreases; this whilst the maximum $\Delta T$ inside the BL motor seems to remain almost constant for few seconds before it starts to linearly decrease. These differences in $\Delta T$ can be explained by considering two factors: the first is the sudden reduction of the convective heat flow coefficient that affects mainly the external components of the engine, the second one is related to the direction of the heat flow from inside (filament winding) to outside.

The influence of the PWM provided to the BL motor is shown in Fig. 4 in terms of $\Delta T$ evaluated inside the black rectangle of Fig. 3a; three different values of PWM are considered.

As expected, the mean $\Delta T$ value increases with the electrical power provided in input to the electric motor.

## 4 Conclusions

The infrared Camera Flir T650sc has been used to perform some preliminary monitoring tests on a small BL motor, which is used in small UASs vehicles, for several working conditions. The goal of the experimental activity was to verify whether infrared thermography could be applied as monitoring tool for predictive maintenance purposes of these particular devices, in order to prevent sudden and catastrophic failures during specific tasks [9, 10]. The experimental campaign highlighted some key aspects, which must be taken into account in order to define best practices and procedures for successfully monitoring the thermal behaviour of some properly operating Turnigy C2830-1050 BL motors.

The obtained results seem promising but other tests should be performed by considering: different electrical engine models, more working conditions and other infrared camera models.

**Acknowledgements** The authors express their sincere gratitude to Mr. Alberto Greco for the technical support.

# References

1. Papa U (2018) Embedded platforms for UAS landing path and obstacle detection: integration and development of unmanned aircraft systems, vol 136. Springer, Berlin
2. Madonna V, Giangrande P, Galea M (2018) Electrical power generation in aircraft: review, challenges, and opportunities. IEEE Trans Transp Electrifications 4(3):646–659
3. Vollmer M, Möllmann K (2018) Infrared thermal imaging; fundamentals, research and applications, 2nd edn. Wiley-VCH Verlag GmbH & Co. KGaA. ISBN: 978-3-527-41351-5
4. Meola C, Boccardi S, Carlomagno GM (2016) Infrared thermography in the evaluation of aerospace composite materials. Woodhead Publishing Print Book, 180 p. ISBN: 9781782421719
5. Meola C, Boccardi S, Carlomagno GM (2018) A quantitative approach to retrieve delamination extension from thermal images recorded during impact tests. NDT & E Int 100:142–152
6. Irace A (2012) Infrared thermography application to functional and failure analysis of electron devices and circuits. Microelectron Reliab 52:2019–2023
7. Singh G, Anil Kumar TCh, Naikan VNA (2016) Induction motor inter turn fault detection using infrared thermographic analysis. Infrared Phys Technol 2016(77):277–282
8. (2008) Standard for infrared inspection of electrical systems & rotating equipment. Infrared inspection Institute
9. Ippolito CA (2019) Dynamic ground risk mitigating flight control for autonomous small UAS in urban environments. AIAA Scitech 2019 Forum, 7–11 Jan 2019, San Diego, California
10. Petritoli E, Leccese F, Ciani L (2018) Reliability and maintenance analysis of unmanned aerial vehicles. Sensors 2018(18):3171

# Portable Fluorescence Sensor for Organic Contaminants and Cyanobacterial Detection in Waters

Gianluca Persichetti⊙, Doriane Combot, Pablo Pelissier, Saverio Savio, Genni Testa⊙, Roberta Congestri⊙, Laurent Labbe, and Romeo Bernini⊙

**Abstract** A portable sensor for fluorescence spectroscopy of liquids has been developed. This sensor exploits the autofluorescence produced by organic chemical compounds when exposed to ultraviolet or visible light. In particular, the autofluorescence produced by pigments present in cyanobacteria and microalgae allow their in vivo detection. Liquid jet waveguide approach has been chosen in order to provide efficient fluorescence light collection. A mini-spectrophotometer is used as a detector and a mini-PC manages all the instrumentation. The whole sensor is housed inside a weather-proof plastic enclosure which allows it to be transported as hand luggage. The sensor can be equipped with three different LED sources that can be easily replaced according to the needs due to the type of substances intended to detect. Several sources at different wavelengths have been taken into consideration and characterized for this purpose. The excitation at multiple wavelengths provides the possibility of discrimination between different substances or phytoplankton groups. Preliminary measurements, using excitation wavelength at 275, 405 and 590 nm, show the possibility of detecting the presence of cyanobacteria in water collected at different pickup points in an aquaponics plant.

**Keywords** Online sensor · Cyanobacteria detection · In vivo fluorescence

G. Persichetti (✉) · G. Testa · R. Bernini
Institute for Electromagnetic Sensing of the Environment (IREA) National Research Council (CNR), 80124 Naples, Italy
e-mail: persichetti.g@irea.cnr.it

S. Savio · R. Congestri
Department of Biology, University of Rome 'Tor Vergata', Via Cracovia 1, 00133 Rome, Italy

D. Combot · P. Pelissier · L. Labbe
INRA, UE 0937 PEIMA (Experimental Fish Farming INRA of Monts d'Arrée), 29450 Sizun, France

© The Author(s), under exclusive license to Springer Nature Switzerland AG 2021       77
G. Di Francia and C. Di Natale (eds.), *Sensors and Microsystems*,
Lecture Notes in Electrical Engineering 753,
https://doi.org/10.1007/978-3-030-69551-4_11

# 1 Introduction

The last few decades have seen a rapid intensification of algal bloom phenomena in surface water and increased frequency in the occurrence of organic contaminants in potable water and wastewater. All these phenomena require rapid detection in order to adopt an appropriate and effective intervention. To date, laboratory tests for the detection of water contaminants can usually take 2–5 days. This time may be insufficient to take effective countermeasures. Among possible alternative strategies to laboratory tests, fluorescence spectroscopy is considered a sensitive and reliable method for the detection of chemical pollutants [1]. In addition to the interest resulting from the detection of chemicals, fluorescence approach is a widespread method for detection of cyanobacteria in aquatic media. This is because fluorescence is a very suitable technique to be implemented for in situ sensors allowing to discriminate and quantify photosynthetic microorganisms in their natural environment, sometimes providing an estimate of their biomass. Furthermore, it is possible to detect the presence of specific taxonomic groups through their specific spectral signature, e.g. specific autofluorescence signal given by specific pigment patterns [2]. In this work, we have combined the advantages of fluorescence spectroscopy with that of a liquid jet waveguide [1] to obtain a sensor in which this latter approach is exploited to improve the collection of the fluorescence signal. More specifically, a portable system based on autofluorescence in a water jet waveguide that exploits visible and ultraviolet excitation at three different wavelengths has been developed. This research is part of the development of an integrated monitoring system to control the presence of microbial hazards in the waters of an aquaponic system, which can affect food safety.

# 2 Sensor Setup

The developed sensor, suitable for in situ measurements, is based on the water jet waveguide approach [1]. This method, has been already used and described for Raman [3] and fluorescence spectroscopy [1], it allows high efficiency of fluorescence collection thanks to the total internal reflection (TIR) that occurs in a jet of water, which acts as a liquid optical fiber, as it is shown in Fig. 1a.

The excitation of the samples is achieved through the use of three LEDs emitting at different wavelengths. A modular approach was used in the sensor design. In particular, it was decided to use optomechanical components that allow a rapid replacement of the LED sources. In this way, the sensor can be easily re-modulated for the detection of specific substances according to the needs. In particular, LEDs emitting at 265, 275, 365, 405 and 590 nm were tested and used. But in the following we will refer almost exclusively to applications related to excitation wavelength at 275 nm, 405 and 590. The sample to be analyzed is injected by means of a mini-pump into a steel capillary which has an inner diameter of about 1 mm. The sensor can work both online and with a limited quantity of sample, taking the solution from

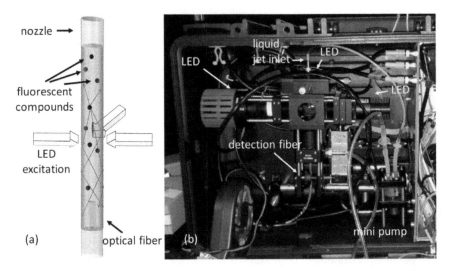

**Fig. 1** Liquid jet waveguide approach scheme (**a**) Optical part of the portable sensor: liquid jet production system (micropump, water outlet nozzle, water outlet tray which can be connected again to the micropump for recirculation) the three excitation LEDs with relative optical filters. The optical fiber that collects the signal is connected to a mini spectrophotometer (**b**)

a bottle through a recirculation system activated by the mini-pump. In this work, the latter method was used. When the sample exits the capillary, it gives rise to a jet (i.e. a regular cylinder of liquid) that acts as a liquid waveguide due to the TIR at the liquid–air interface. Enabled one at a time, each of the three LEDs excites the fluorescence of the organic substances present in the sample. Then, an optical fiber, optically connected with the liquid jet, delivers the fluorescence signal towards a mini-spectrophotometer used for the detection. An internal mini-PC manages a mini-spectrophotometer, a motorized selector for detection filters and LED drivers via USB. Software written in MATLAB code is used both for the management of the different elements of the setup and for data analysis.

The entire experimental setup is housed inside a plastic transport container measuring 56 cm × 32 cm × 27 cm. The optical part of the portable sensor is shown in Fig. 1b. The LED that emits at 275 nm is intended to excite the so-called tryptophan-like fluorescence (TLF). The TLF peak (centered at 340 nm) is so named because it reflects the presence of compounds that have similar fluorescence charac-teristics as the amino acid tryptophan. Several studies demonstrate the use of TLF detection as indicator of microbial contaminations in water [4, 5]. Moreover, although it is not the most suitable wavelength, an LED, emitting at 275 nm, allowed to excite the fluorescence of chlorophylls, phycobiliproteins, such as phycocyanin and phyco-erythrin, but also a set of cellular metabolites that exhibit autofluorescence such as nicotinamide adenine dinucleotide (NADH or NAD(P)H) and flavins. The LED that emits at 405 nm is intended to efficiently excite chlorophylls, whereas the LED emit-ting at 590 nm has the purpose of selectively exciting phycocyanin. The selective detection of this pigment, being specific to cyanobacteria, allows a differentiation of cyanobacteria from other microalgae [6]. The sensor will be characterized in order to obtain the detection limits of common organic pollutants and microorganisms. In addition, the sensor will be used for the detection of cyanobacteria in an aquaponics system of the INRA-PEIMA experimental fish farm that is located in Sizun (France).

## 3   Fluorescence Measurements

Due to its operating principle, the sensor is capable of detecting a wide range of fluorescent organic compounds in water. However, in this work we focused our attention in cyanobacteria detection.

Thanks to the large collection of algal strains available at the Laboratory of Biology of Algae, Department of Biology, University of Rome "Tor Vergata", it has been possible to examine a large number of of cyanobacteria and microalgae with different pigment patterns, size and cellular organizations, either isolated in freshwater or from brackish, dystrophic environments. Only as a representative example, four different photosynthetic strains are shown in Fig. 2. As pointed out in Sect. 2, the LED emitting at 275 nm gives an overview of many fluorescent compounds present in the photosynthetic microorganisms. From Fig. 2 it is further noted that the peak of phycocyanin is particularly evident in the spectrum of *Microcystis aeruginosa* SAG 17.85 (as well as in the other cyanobacteria examined). This aspect can be

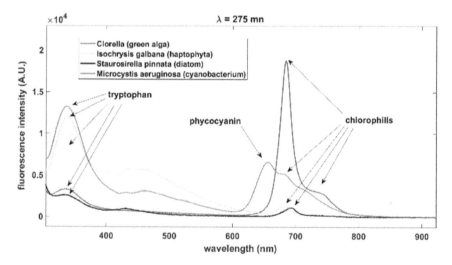

**Fig. 2** Example of in vivo fluorescence spectra of typical photosynthetic microorganisms acquired with the portable sensor (excitation wavelength $\lambda = 275$ nm). Some of the peaks attributable to specific pigments useful for discrimination are indicated with arrows

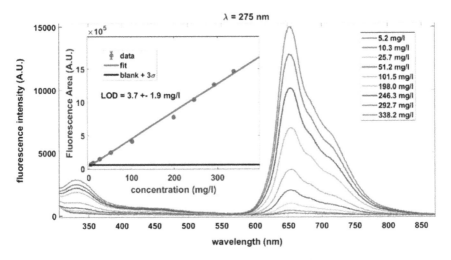

**Fig. 3** Sensor calibration with *Trichormus variabilis* at different concentrations (excitation wavelength $\lambda = 275$ nm). In the inset the calibration curve of the sensor performing a LOD = 3.7 mg/l

exploited for the determination of the presence of cyanobacteria [6]. From these fluorescence spectra (and from others acquired with additional excitation wavelengths) it is possible to deduce the presence of characteristic peaks that will be useful for building a database for future discrimination applications of our sensor. In order to test the ability of our portable device in detecting cyanobacteria at low concentration we considered a sensor calibration with the cyanobacterium *Trichormus variabilis*, cultured in BG11$_0$ medium. The measurements have been performed with sample concentrations spanning around two order of magnitude. Sample concentration have been determined by measuring their dry weight after measurements. Examples of typical measurements performed for calibration are shown in Fig. 3, where samples are excited at 275 nm, in the inset is reported the calibration curve and the corresponding limit of detection (LOD). Although there are some differences depending on the LED used for calibration, the LODs achieved are typically in the order of a few mg/l. The best result was obtained with the LED at 405 nm (LOD = 1.2 $\pm$ 0.5 mg/l) while the highest LOD was obtained with the LED at 265 nm (LOD = 20.8 $\pm$ 10.6 mg/l). The difference between the measured values, as well as the different efficiency of the excited pigments, at those specific wavelengths, is mainly due to the different emission powers that can be reached by the LEDs.

Although the portable sensor is not intended for this purpose, the results are competitive with values between 1 and 5 mg/l such as those of commercial handheld devices for cell counting for estimating biomass [7]. The sensor has been applied for the control of the presence of microbial contamination (in particular due to cyanobacteria) in the waters of the aquaponic farm of INRA-PEIMA located in Sizun (France). For this specific purpose, the most appropriate choice, determined after a measurement campaign carried out over about two years, was to equip the portable sensor,

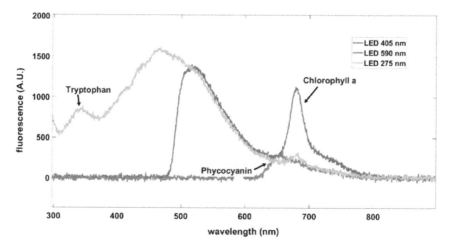

**Fig. 4** Example of detection of waters from the aquaponic farm of INRA-PEIMA. The presence of cyanobacteria is signaled by the presence of the phycocyanin peak

with LEDs emitting at wavelengths of 275, 405 and 590 nm. These investigations were carried out by taking samples in different points of the experimental fish farm of INRA-PEIMA and sent both to our laboratory to be analyzed by our sensor and to an external laboratory to be analyzed and counted by means of counting chambers method. An example of detection of waters from the aquaponic farm is shown in Fig. 4, where the assignments of the most relevant fluorescence peaks are also highlighted. In the specific sample, shown in Fig. 4, a cyanobacteria concentration of 7508 cells/ml has been measured by counting cells.

## 4  Conclusions

A sensor potentially capable of detecting the presence of a wide range of organic substances in water has been developed. The presence of a mini PC on board for an immediate analysis of the measurements and the management of the instrumentation allow an easy use for in situ applications while the jet waveguide approach allows in vivo analysis of biological samples. After a series of tests and in vivo measurements, the portable sensor has been applied to the monitoring of waters potentially subject to the presence of cyanobacteria. Each of the chosen excitation wavelengths has a specific role. The excitation at 275 nm provides the "tryptophan like" fluorescence useful for an estimate of the whole biomass, allowing for an overview of the organic substances present in the sample. The wavelength of 405 nm effectively excites chlorophyll *a* highlighting the presence of photosynthetic microorganisms.

The excitation at 590 nm excites phycocyanin in a fairly selective way and since the latter is a characteristic accessory pigment of cyanobacteria, it provides a clear sign of their presence.

**Acknowledgements** This research has been partially supported by the ERA-NET Cofund WaterWorks 2015 project SMARTECOPONICS, "On-Site Microbial Sensing For Minimising Environmental Risks From Aquaponics To Human Health".

# References

1. Persichetti G, Testa G, Bernini R (2013) High sensitivity UV fluorescence spectroscopy based on an optofluidic jet waveguide. Opt Express 21:24219–24230
2. MacIntyre HL, Lawrenz E, Richardson TL (2010) Taxonomic discrimination of phytoplankton by spectral fluorescence. In: Suggett D, Prášil O, Borowitzka M (eds) Chlorophyll a fluorescence in aquatic sciences: methods and applications. Developments in applied phycology, vol 4. Springer, Dordrecht
3. Persichetti, Bernini R (2016) Water monitoring by optofluidic Raman spectroscopy for in situ applications. Talanta 155:145–152
4. Nowicki S, Lapworth DJ, Ward JST, Thomson P, Charles K (2019) Tryptophan-like fluorescence as a measure of microbial contamination risk in groundwater. Sci Total Environ 646:782–791.
5. Cumberland S, Bridgeman J, Baker A, Sterling M, Ward D (2012) Fluorescence spectroscopy as a tool for determining microbial quality in potable water applications. Environ Technol 33(6):687–693
6. Gregor J, Maršálek B (2005) A simple in vivo fluorescence method for the selective detection and quantification of freshwater cyanobacteria and eukaryotic algae. Acta Hydrochim Hydrobiol 33:142–148
7. tipbiosystems.com    https://tipbiosystems.com/wp-content/uploads/2020/05/AN50-Spirulina-biomass_2018_02_02.pdf. Last accessed 2020/05/22.

# Augmented Reality (AR) and Brain-Computer Interface (BCI): Two Enabling Technologies for Empowering the Fruition of Sensor Data in the 4.0 Era

**Annarita Tedesco, Dominique Dallet, and Pasquale Arpaia**

**Abstract** The role of sensors is particularly relevant in the current 4.0 Era. In fact, the 4.0 paradigm thrives on data provided by sensors. On such basis, this work addresses the use of brain computer interface (BCI) and augmented reality (AR) as User's input/output interfaces with sensors. Following a User-centered approach to 4.0 Era applications, the aforementioned technologies are expected to facilitate human interaction with the surrounding sensory environment in this new Era. For the sake of example, this work presents an overview of three experimental case studies addressed by the Authors: two in the industrial field and one in the medical field. Although experimented in specific field, the obtained results and the supporting considerations can be readily extended to any other 4.0 Era application field.

**Keywords** Augmented reality · Brain computer interface · Internet of Things · 4.0 Era · Sensors

## 1 Introduction

The role of sensors is undoubtedly relevant to the current 4.0 Era. In fact, any enabling technologies of the 4.0 paradigm thrives on raw measurement data provided by sensors. The internet of things (IoT) has contributed in boosting the pervasiveness of sensors in every field. Nowadays, however, the challenge is not only to monitor a physical quantity, but also to facilitate and make more effective the User's fruition of the information made available by the sensors.

An analysis of the operating context suggests the importance of adopting a User-centered approach, to improve usability and User experience in the 4.0 Era applications [1]. Especially, in Industry 4.0 applications, humans are becoming a composite

A. Tedesco (✉) · D. Dallet
IMS Laboratory, Bordeaux INP, University of Bordeaux, Bordeaux, France
e-mail: annarita.tedesco@unina.it

P. Arpaia
Department of Information Technology and Electrical Engineering, University of Naples Federico II, Naples, Italy

G. Di Francia and C. Di Natale (eds.), *Sensors and Microsystems*,
Lecture Notes in Electrical Engineering 753,
https://doi.org/10.1007/978-3-030-69551-4_12

factor in a highly automated system; therefore, to improve the collaboration between men and the surrounding 4.0 Era context, it is necessary to enhance the human–machine interaction. This is particularly important not only in the industrial sector, but also in the medical one: in fact, in both scenarios, the overall effectiveness of any 4.0 process is intrinsically related to the performance of the human actor involved.

A promising communication channel is offered by the combination of brain computer interface (BCI) and augmented reality (AR). AR is the technology that increases reality perception by mixing physical objects with digital information, and providing the necessary feedback to the User [2]. Instead, brain computer interface (BCI) has the potential to become the "ultimate interface", through which thoughts can become acts [3, 4].

Starting from these considerations, this work outlines the main results of the research activities related to the use of AR and BCI as sensor-to-human interfacing solutions, in the industrial and medical fields.

First, the general diagram of the envisaged human/sensor interaction is described. Successively, three experimental cases that implement the proposed human-sensor interaction model are reported.

In particular, the first case study relates to the adoption of both AR and BCI for hands-free interrogation of remote sensors. The second experimental case reports on the use of AR for assisting an industrial assembly task, and the analysis of the operator's performance in his interaction with the working environment [5]. Finally, the third case reports on an ongoing research activity related to the use of AR in the operating room during surgical procedures: in particular, for an effective fruition by the anesthetist of the data, i.e. the patient's vitals, coming from the monitoring instrumentation [6].

It should be mentioned that, although the reported cases refer to specific application scenarios, the results and the considerations can be readily extended to any other application field.

## 2 Schematization of the Proposed User-Centered Design Approach

Figure 1 shows a schematization of a user-centered, combined BCI/ AR interface, which is at the base of the approach described in this paper.

A *wireless sensor network* measures the physical quantities of interest from the system to be monitored.

The User wears a BCI/AR interface, which includes AR glasses; noninvasive, dry electrodes for the electroencephalogram (EEG); and a wearable computer for data processing. Basically, the BCI acts as an input interface, whereas the AR acts as an output interface.

BCI allows the User to select among different possible choices (i.e., alternative inputs). In one possible configuration, a BCI may exploit steady state visually evoked

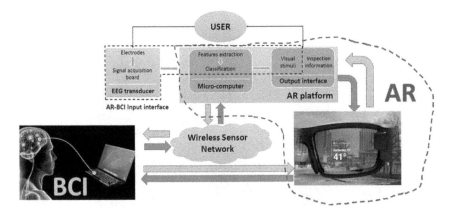

**Fig. 1** Schematization of the design of user-centered BCI/ AR interface

potentials (SSVEPs). Visual stimuli are shown on the display of the AR glasses worn by the operator: each visual stimulus corresponds to a different possible input and flickers at a different frequency. When the User wants to select the input, he stares at the associated flickering signal. This naturally generates ("evokes") electrical activity in the operator's brain, at the same (or multiples of the) frequency of the visual flickering stimulus [4, 7]. The brain signal is collected through a portable EEG with wearable dry electrodes, and it is processed by a wearable microcomputer. The real-time, automated analysis of the signal features of the brain signal discloses the selection that was made by the User.

If we suppose that the User's selection regards the visualization of the output from a number of sensors, then the selected sensor's output will be shown to the User on the display of the AR glasses.

## 3 Experimental and Results

### 3.1 Case #1

Figure 2 shows the practical implementation of the system sketched in Fig. 1. In particular, this application refers to the hands-free interrogation of remote sensors (in this experiment, the sensor output was emulated). The picture on the left in Fig. 2 shows the user wearing the AR glasses (Epson Moverio BT350), the dry noninvasive EEG electrodes, and a wearable computer [4].

The picture on the right of Fig. 2 shows what the user see through his AR glasses. In practice, the writings "humidity" and "temperature" flicker at different frequencies. The two writings are associated with the output of the corresponding sensors.

 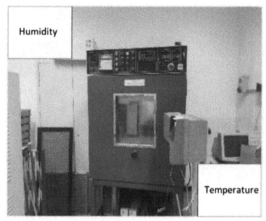

**Fig. 2** On the left, picture of a User wearing AR glasses and the BCI. On the right, picture of what the User sees through the AR glasses. The writings "Humidity" and "Temperature" flicker at different frequencies [4]

When the User stares for approximately one second at either the writings, the corresponding features are extracted from the acquired EEG signal. This processing takes less than one second. The automated analysis of the EEG signal allows to identify which selection was made by the User. Finally, the measurement information from the selected sensor is displayed through the AR glasses.

## 3.2 Case #2

The second case relates to the use of AR as an interface to provide visual instructions to an operator in a typical industrial application, namely the assembly of a product.

The case study consists of an assembly task in which the operator is required to assemble a prototype made with Legos, for which the assembly instructions are available in either paper or virtual-aided forms. Figure 3 (on the left) shows a picture of the assembly station and of an operator carrying out the assembly task following AR instructions. On the right of Fig. 3, the example of an instruction image shown to the operator through the AR glasses [5].

Comparative experimental tests were carried out on two groups of volunteers, who had to complete the task. One group followed paper instructions, whereas the other followed instructions administered through AR glasses.

The goal of this research activity was to assess objectively the beneficial effects introduced by the adoption of AR. To this end, two figures of merit were considered in the measurement: (1) the overall processing time for the operator to complete the task; (2) the number of mistakes the operator made during the execution of the task.

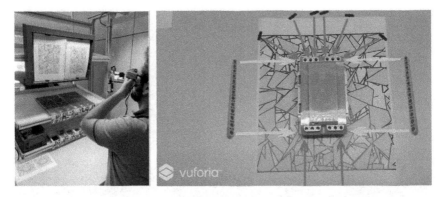

**Fig. 3** On the left, there is a picture of the assembly table and of an operator carrying out the assembly task following AR instructions. On the right, the example of an instruction image shown to the operator through the AR glasses [5]

Results obtained in terms of operators' learning curves showed that the group of people who used AR instructions slightly outperformed the group that used paper instructions. The most notable observation, however, was that the group with AR instructions made almost no errors in the assembly procedure. This confirmed the positive effect related to the use of AR as a User interface with real world. This beneficial effect is expected to be more marked for more complex assembly tasks.

### 3.3 Case #3

The third case study reported in this work regards to an AR-based system for monitoring patient's vitals in the operating room (OR). An optical see-through (OST) headset, worn by the anesthetist or by the surgeon's assistant, shows in real-time the patient's vitals acquired from the electro-medical equipment available in the OR [6]. A dedicated application was developed to allow a hands-free fruition of the AR content. Using AR glasses as a User interface allows the anesthetist to maintain a higher level of concentration on the task at hand. Experimental tests were carried out acquiring the vital parameters from two pieces of equipment typically available in the OR, namely a ventilator and a monitor. Figure 4 shows the AR visualization of some vital parameters made available on the AR glasses. Because of the stringent requirement of the applications, experimental tests were carried out to assess the transmission error rate and the latency: results, so far, have demonstrated the reliability of the proposed AR-based monitoring system.

Current research is focused on including additional "augmented" information for the User: (1) selected information from the patient's electronic medical record; and (2) real-time alert information on the health status of the patient, based on the processing made through dedicated predictive algorithms.

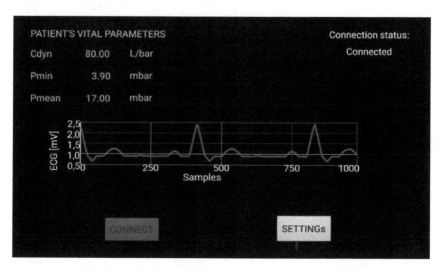

**Fig. 4** Example of the visualization of the patient's vitals acquired from the operating room instrumentation, as shown through the AR glasses [6]

## 4 Conclusion

In this work, the use of BCI and AR as output/input interfaces between the User and the sensors in a 4.0 Era scenario is proposed. These two enabling technologies can be used (separately or combined with each other), to guarantee an effective fruition of the information generated by the sensors. The common thread in all the experimental cases reported herein is the fact that all the applications are designed following a user-centered approach. This was done starting from the consideration that, also in the 4.0 Era, the effectiveness of any application that involves humans may be limited by humans' performance, especially when they do not "blend in" and feel comfortable with the surrounding technologies. Current research activity is being dedicated to further improving the effectiveness of the proposed BCI/AR interfaces. This is done both in terms of usability (e.g., by identifying the optimal frequency values of the flickering visual signals in BCI, not to discomfort the user), and increase of the informative content made available to the User (e.g., by adding additional information for the User, as described in the medical application).

## References

1. Pfeiffer A et al (2016) Empowering user interfaces for Industrie 4.0. Proc IEEE 104:5:986–996.
2. Paelke V (2014) Augmented reality in the smart factory: supporting workers in an Industry 4.0. environment. In: Proceedings of the 2014 IEEE emerging technology and factory automation (ETFA). IEEE, 2014.

3. Wang Y et al (2006) A practical VEP-based brain-computer interface. IEEE Trans Neural Syst Rehabil Eng 14(2):234–240
4. Angrisani L et al (2020) A wearable brain-computer interface instrument for augmented reality-based inspection in Industry 4.0. IEEE Trans Instrum Meas 69(4):1530–1539. https://doi.org/10.1109/TIM.2019.2914712
5. Bonavolontà F et al (2020) Measuring worker's performance in augmented reality-assisted Industry 4.0 procedures. In: 2020 IEEE international instrumentation and measurement technology conference (I2MTC), Dubrovnik, Croatia, 2020, pp 1–6, https://doi.org/10.1109/I2MTC43012.2020.9129320
6. Arpaia P et al A health 4.0 integrated system for monitoring and predicting patient's health during surgical procedures. In: 2020 IEEE international instrumentation and measurement technology conference (I2MTC), Dubrovnik, Croatia, 2020, pp 1–6. https://doi.org/10.1109/I2MTC43012.2020.9128840
7. Sohn RH et al (2010) SSVEP-based functional electrical stimulation system for motor control of patients with spinal cord injury. In: Lim CT, Goh JCH (eds) 6th world congress of biomechanics (WCB 2010). Aug 1–6, 2010 Singapore. IFMBE Proceedings, vol 31. Springer, Berlin, Heidelberg

# Hybrid Organic/Inorganic Nanomaterials for Biochemical Sensing

Ilaria Rea, Principia Dardano, Rosalba Moretta, Chiara Schiattarella, Monica Terracciano, Maurizio Casalino, Mariano Gioffrè, Teresa Crisci, Giovanna Chianese, Chiara Tramontano, Nicola Borbone, and Luca De Stefano

**Abstract** In this paper, different nanostructured semiconductors with advanced properties are explored for the realization of both optical and electrical biosensors for DNA detection. A hybrid sensor constituted by graphene oxide (GO) covalently grafted on a porous silicon (PSi) matrix is realized. A peptide nucleic acid (PNA) probe, able to recognize its complementary DNA (c-DNA) sequence, is immobilized on the surface of PSi/GO device for label-free optical sensing. Electrical sensing of DNA is also demonstrated using a Zinc Oxide Nanowires (ZnONWs) sensor functionalized with PNA probe; the I–V characteristic of the device depends on the c-DNA concentration under analysis.

**Keywords** Nanomaterial · Semiconductor · PNA probe · Bioconjugation · Biochemical sensing

## 1 Introduction

Biosensors are analytical devices constituted by a biological element, generally defined as probe, immobilized on a transduction system, converting a molecular interaction in a measurable output signal. In the last decades, the growing prevalence of diseases and the increasing demand for their rapid and reliable detection fueled the biosensors market. Biosensors industry includes two categories of bioanalytical devices: traditional practices in clinical laboratories, requiring trained personnel and characterized by a long and expensive samples processing; cheap, portable and easy-to-use point-of-care (POC) devices for analyses in both clinical and nonclinical ambients. The successful commercialization of POC glucose tests (e.g. FreeStyle Lite®, Abbott Inc.) favored the development of other innovative portable diagnostic

I. Rea (✉) · P. Dardano · R. Moretta · C. Schiattarella · M. Casalino · M. Gioffrè · T. Crisci · G. Chianese · C. Tramontano · L. De Stefano
Institute of Applied Sciences and Intelligent Systems, National Research Council, Naples, Italy
e-mail: ilaria.rea@cnr.it

M. Terracciano · N. Borbone
Department of Pharmacy, University of Naples Federico II, Naples, Italy

© The Author(s), under exclusive license to Springer Nature Switzerland AG 2021
G. Di Francia and C. Di Natale (eds.), *Sensors and Microsystems*,
Lecture Notes in Electrical Engineering 753,
https://doi.org/10.1007/978-3-030-69551-4_13

devices that contribute, today, to approximately 57% of overall biosensors market [1]. On the other hand, the recent advances in nanotechnology are helping to address challenges in conventional biosensing such as high sensitivity, selectivity, minimal sample volume, low costs, fast detection. Nanomaterials can offer advanced properties compared to the corresponding bulk materials, such as unique physico-chemical features and a very high specific surface area that assures an increase of sensitivity for the detection of target analytes.

In this work, different nanostructured platforms were explored for the realization of both optical and electrical portable biosensors. Silicon and its related materials (i.e. silicon oxide, silicon nitrides and porous silicon (PSi)), the semiconductors mainly used in the microelectronic industry, were extensively exploited in biosensors fabrication [2]. Other nanostructured oxides, such as Graphene Oxide (GO) and Zinc Oxide Nanowires (ZnONWs) were also investigated as sensing platforms due to their peculiar properties [3–6].

The devices were obtained by merging conventional microelectronic techniques with liquid synthesis processes and surface modification strategies. Since a valid surface functionalization strategy is crucial in the development of biosensors in order to correctly immobilize the bioprobes on the material surfaces, several chemical processes were explored in this work. In particular, strategies based on silane (e.g. APTES, APDMES) grafting and hydrosilylation were used as treatments to passivate and/or functionalize nanostructured surfaces with biomolecules such as DNA or peptide nucleic acid (PNA) [7, 8].

## 2 Experimental

### 2.1 Chemicals

Hydrofluoric acid (HF), undecylenic acid (UDA), N-(3-Dimethylaminopropyl)-N′-ethylcarbodiimide hydrochloride (EDC), N-hydroxysuccinimide (NHS), MES hydrate, Dimethyl sulfoxide (DMSO), tert-Butyloxycarbonyl-NH-PEG-Amine (BOC-NH-PEG-NH$_2$), trifluoroacetic acid (TFA), chloroform, tetrahydrofuran, (3-Aminopropyl)triethoxysilane (APTES), toluene, bis(sulfosuccinimidyl)suberate (BS$^3$) were purchased from Sigma Aldrich (St. Louis, MO, USA), Graphene oxide (GO) nanosheets were purchased from Biotool.com (Houston, TX, USA) as a batch of 2 mg/ml in water with a nominal size sheets between 50 and 200 nm.

### 2.2 Fabrication of the PNA-Based PSi/GO Optical Sensor

PNA-based PSi/GO hybrid device was realized following the scheme reported in Fig. 1. PSi structure was fabricated by electrochemical etching of n-type crys-

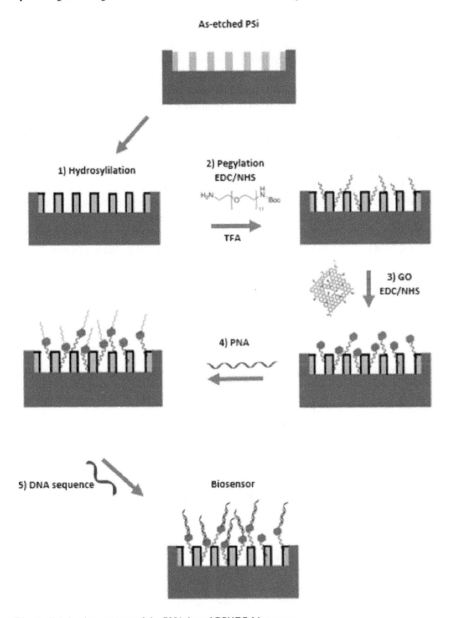

**Fig. 1** Fabrication process of the PNA-based PSi/GO biosensor

talline silicon (0.01–0.02 $\Omega$ cm resistivity, $\langle 100 \rangle$ orientation, 500 $\mu$m thickness) in HF (5% in weight)/ethanol solution at room temperature. A current density of 20 mA cm$^{-2}$ for 90 s was applied to obtain a single layer of PSi with 61% in porosity ($n_{PSi} = 1.83$ at $\lambda = 1.2$ $\mu$m), thickness L of 2.1 $\mu$m and pore dimension included between 50 and 250 nm. The as-etched PSi was placed in a Schlenk tube containing deoxygenate neat UDA (99% v/v). The reaction was conducted at 110 °C for 18 h in argon. The hydrosilylated-PSi was extensively washed in chloroform and tetrahydrofuran [9]. The carboxyl acid groups of UDA were activated by EDC/NHS (0.005 M in 0.1 M MES buffer) for 90 min at RT. Afterwards, a solution containing BOC-NH-PEG-NH$_2$ (0.004 M, overnight at 4 °C) was used to completely recover the PSi chip. A solution of TFA (95% v/v, 90 min, RT) was used to deprotect the amino group of the PEG molecule. The PSi chip was washed with deionized water. The GO was activated by EDC/NHS and added to the PSi chip, to allow the covalent bond. The GO-PSi device was exposed to 100 $\mu$M of PNA (COOH-KKTCGTGGCTCGGGNHCOCH$_3$) in presence of EDC/NHS and washed in deionized water. Complementary (5′-AGGAGAGCACCGAGCCCCTGAG-3′) and non-complementary (5′-CCTTTTTTTTTT-3′) DNA sequence were incubated on the chip for two hours. The sample was gently rinsed in deionized water to remove the unhybridized target.

## 2.3 Spectroscopic Reflectometry

The reflectivity of PSi/GO device was measured at normal incidence by means of a Y optical reflection probe (Avantes), connected to a white light source and to an optical spectrum analyzer (Ando, AQ6315B).

## 2.4 Fabrication of the PNA-Based ZnONWs Electrical Sensor

The ZnONWs electrical sensor was realized using the fabrication process described in the following. A uniform ZnO seed layer, 150 nm thick, was deposited on a thermally oxidized silicon substrate using a radio frequency (RF) magnetron sputtering equipment with a 99.999% pure ceramic ZnO target. ZnONWs were then growth on the sputtered ZnO thin film by hydrothermal synthesis, immersing the substrate in a solution obtained by dissolving 0.5 M of hexamethylenetetramine ($C_6H_{12}N_4$) and the $Zn^{2+}$ salt ($Zn(NO_3)_2$), which acts as a precursor, in deionized water. The hydrothermal synthesis was performed at a temperature of 90 °C for 4 h. Gold interdigitated electrodes were deposited on the ZnONWs through an iron shadow mask. The device was finally treated by rapid thermal annealing at 400 °C for 5 min. The ZnONWs surface was functionalized in order to immobilize PNA probe, by applying the silane chemistry. Briefly, naturally hydrolyzed ZnONWs were treated with a solution of 5% APTES in toluene anhydrous for 30 min at room temperature, cured on heater at 100 °C for 10 min. Amino-modified ZnONWs were then treated by cross-linker BS$^3$ 1.7 mM in PBS 1X pH 7.4 at 4 °C for 3 h. The sulfo-NHS-terminated NWs were then incubated at 4 °C overnight with 300 μM PNA dispersed in deionized water. Increasing concentrations (25–200 μM) of complementary DNA were drop-deposited on the PNA-modified ZnONWs sensor for biorecognition monitoring.

## 2.5 Electrical Characterization

The I−V characteristics of the ZnONWs biosensor were acquired using a Source Meter Agilent B2902A. Samples were accommodated on the holder of a probe station and electrodes connected by two XYZ micromanipulators (Micromanipulator, 450/360MT-6 and 550/360MT-6 series) provided of micrometric tips.

## 3 Results and Discussion

PSi/GO sensor is a nanostructured hybrid device characterized by a large specific surface area useful for an efficient sensing. Moreover, the presence of GO makes the device COOH-terminated and allows the covalent immobilization of biomolecules containing –NH$_2$ groups. The probe used for the biosensor realization is a peptide nucleic acid (PNA), a nucleic acid with a peptide backbone that does not contain any charges. The employment of PNA instead of DNA as a probe for complementary-DNA (c-DNA) detection, enables a high hybridization efficiency due to its non-ionic backbone. The interaction between PNA and c-DNA at different concentrations was monitored by label-free optical measurements based on the reflectance spectrum analysis (Fig. 2). Red-shift variations were registered at increasing c-DNA concentrations; the phenomenon is due to an increase of the average refractive index of the device.

ZnO nanostructures (i.e. nanoribbons, nanorods, nanowires) are interesting transducer materials for biochemical sensing applications due to their large surface over volume ration and very reactive surface. The main technologies used for fabrication of ZnO nano-objects are Vapor–Liquid–Solid growth (VLS), Metal Organic Chemical Vapor Deposition (MOCVD) and High Pressure Pulsed Laser Deposition (HP-PLD), requiring complex-equipment [10]. ZnONWs, realized in this work, were obtained by an alternative approach based on the hydrothermal synthesis. Figure 3a shows a photograph of the ZnONWs sensor; in the inset, a SEM image of ZnONWs is reported. ZnONWs appeared as columns perpendicular to the substrate, with a diameter of about 200 nm. Empty spaces of some hundreds of nanometers were observed between adjacent columns of ZnONWs. The current over voltage profile

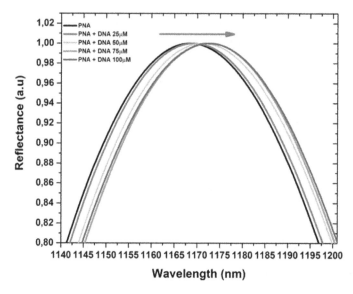

**Fig. 2** Reflectance spectra of PNA-based PSi/GO biosensor after interaction with increasing concentration of c-DNA

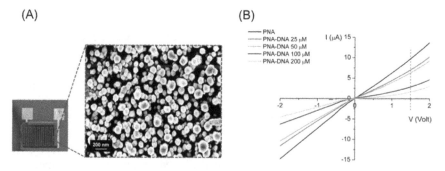

**Fig. 3** **a** A photograph of the ZnONWs electrical sensor; in the inset, a SEM image of ZnONWs grown by hydrothermal synthesis. **b** Current over voltage profiles of ZnONWs sensor functionalized with PNA probe, before and after interaction with increasing concentrations of complementary DNA

of the electrical biosensor (i.e. ZnONWs sensor functionalized with PNA probe) was acquired after the exposure to increasing concentrations of c-DNA in the range included between 25 and 200 μM. The results reported in Fig. 3b shown that the presence of c-DNA induced a linear increase of the sensor resistance at 1.5 V.

# 4 Conclusion

A key revolution in the development of portable and efficient biosensors is the integration of structures at the nanoscale. Due to their large specific surface area and enhanced physico-chemical properties compared to related bulk materials, nanostructured sensing materials opened new horizons for biosensing. The biomolecules immobilization on the nanostructured surfaces can be achieved using proper functionalization procedures that guarantee the biological activity of biomolecule. In this work, optical and electrical sensors realized using nanostructured semiconductors are described. The devices were functionalized with PNA, a nucleic acid able to recognize c-DNA with high hybridization efficiency due to its non-ionic backbone. The detection of c-DNA concentrations included between 25 and 200 $\mu$M was demonstrated.

# References

1. Mora-Serò I, Fabregat-Santiago F, Deniere B, Bisquert J (2006) Determination of carrier density of ZnO nanowires by electrochemical techniques. Appl Phys Lett 89:203117
2. Terracciano M, De Stefano L, Borbone N, Politi J, Oliviero G, Nici F, Casalino M, Piccialli G, Dardano P, Varra M, Rea I (2016) Solid phase synthesis of a thrombin binding aptamer on macroporous silica for label free optical quantification of thrombin. RSC Adv 6:86762–86769
3. Moretta R, Terracciano M, Dardano P, Casalino M, De Stefano L, Schiattarella C, Rea I (2018) Toward multi-parametric porous silicon transducers based on covalent grafting of graphene oxide for biosensing applications. Front Chem 6:583
4. Rea I, Sansone L, Terracciano M, De Stefano L, Dardano P, Giordano M, Borriello A, Casalino M (2014) Photoluminescence of graphene oxide infiltrated into mesoporous silicon. J Phys Chem C 118:27301–27307
5. Rea I, Casalino M, Terracciano M, Sansone A, Politi J, De Stefano L (2016) Photoluminescence enhancement of graphene oxide emission by infiltration in an aperiodic porous silicon multilayer. Opt Express 24:24413–24421
6. Politi J, Rea I, Dardano P, De Stefano L, Gioffrè M (2015) Versatile synthesis of ZnO nanowires for quantitative optical sensingof molecular biorecognition. Sens Actuators B 220:705–711
7. De Stefano L, Oliviero G, Amato J, Borbone N, Piccialli G, Mayol L, Rendina I, Terracciano M, Rea I (2013) Aminosilane functionalizations of mesoporous oxidized silicon for oligonucleotides synthesis and detection. J R Soc Interface 10:20130160
8. Shabir Q, Webb K, Nadarassan DK, Loni A, Canham LT, Terracciano M, De Stefano L, Rea I (2018) Quantification and reduction of the residual chemical reactivity of passivated biodegradable porous silicon for drug delivery applications. Silicon 10:349–359
9. Ghulinyan M, Gelloz B, Ohta T, Pavesi L, Lockwood DJ, Koshida N (2008) Stabilized porous silicon optical superlattices with controlled surface passivation. Appl Phys Lett 93:061113
10. Okada T, Agung BH, Nakata Y (2004) ZnO nano-rods synthesized by nano-particle- assisted pulsed-laser deposition. Appl Phys A 79(4–6):1417–1419

# Distributed Acoustic Sensor for Liquid Detection Based on Optically Heated CO$^{2+}$-Doped Fibers

A. Coscetta, E. Catalano, E. Cerri, N. Cennamo, L. Zeni, and A. Minardo

**Abstract** In this work, we propose the use of Coherent Optical Time-Domain Reflectometry (C-OTDR) in order to discriminate the medium surrounding one or more Co$^{2+}$-doped fiber sections. The proposed approach can be used to perform vibration measurements along the whole fiber by means of C-OTDR measurements at 850 nm, as well as liquid detection in selected positions by active thermometry at 1550 nm. By pumping the fiber with a modulated optical pump, the induced fiber temperature modulation is monitored through C-OTDR, as the latter is sensitive to optical phase changes occurring along the sensing fiber. As the induced phase modulation (i.e. the C-OTDR signal) is affected by the thermal conductivity of the surrounding medium, the latter can be retrieved from the phase signal. In our tests, a 5 Hz modulation frequency was chosen to discriminate between air, water, and acetone.

**Keywords** Distributed acoustic sensors · Liquid detection · Cobalt-doped fibers

## 1 Introduction

Recently, distributed acoustic sensors have emerged as one of the most promising technologies for distributed measurements of acoustic disturbance and vibrations along an optical fiber [1–3]. The principle of operation relies on Rayleigh scattering in optical fibers, occurring naturally due to inhomogeneities of the fiber itself. When a short optical pulse of coherent light is sent through an optical fiber, the interference between the contributions due to the single scatterers within the pulse leads to an apparently random variation of the backscattered power, which is detected at the pulse launching end as a function of the time. Any external perturbance acting on the fiber, modifies the position of the scatterers in the perturbed region, leading to a change in the backscattered intensity. Tracking these changes among consecutive

A. Coscetta · E. Catalano · E. Cerri · N. Cennamo · L. Zeni · A. Minardo (✉)
Department of Engineering, University of Campania Luigi Vanvitelli, Via Roma 29, 81031 Aversa, Italy
e-mail: aldo.minardo@unicampania.it

© The Author(s), under exclusive license to Springer Nature Switzerland AG 2021
G. Di Francia and C. Di Natale (eds.), *Sensors and Microsystems*,
Lecture Notes in Electrical Engineering 753,
https://doi.org/10.1007/978-3-030-69551-4_14

acquisitions, one may capture any event occurring along the fiber. This technique is commonly referred to as coherent-OTDR (C-OTDR), or in some cases phi-OTDR (the latter is especially used when the phase, rather than intensity, of the backscattered signal is employed for quantitative measurements). The C-OTDR method is simpler and more cost-effective than other distributed sensing technologies, and therefore it is of interest to extend its applicability beyond vibration measurements [4].

In this work, we demonstrate that a conventional C-OTDR setup can be also used to distinguish any external medium surrounding the fiber is some specific points, based on its thermal properties [5]. The only condition is that, in those points, the fiber is subjected to periodical heating through an external source. This heating can be realized by injecting current in a metal-coated fiber, or, more elegantly, by pumping light into the core of a specialty fiber embedding some metal absorbing light and converting it to heat [6].

In this work, we propose the use of short sections of Cobalt-doped fibers absorbing light at 1550 nm optical wavelength. Therefore, our sensing scheme relies on the use of a single-mode optical fiber, along which one or more sections of $Co^{2+}$-doped fibers are spliced. The sensing fiber is interrogated using 850 nm pump pulses for C-OTDR measurements [7], while being optically heated through a pump light at 1550 nm. The heating light is modulated at low frequency (5 Hz), thus inducing a periodic phase modulation in the specialty fiber, which is detected through C-OTDR measurements. The amount of phase modulation is dependant on the external medium, and in particular from its capability to remove heat from the fiber. In out tests, we show that the proposed approach is able to distinguish between air, water and acetone in two separate sections of the sensing fiber.

## 2   Experimental Results

The C-OTDR sensor used for the tests is schematically shown in Fig. 1. The setup implements an intensity-based heterodyne C-OTDR scheme, with the addition of an 850 nm/1550 nm wavelength division multiplexing (WDM) splitter. Heating is performed by a 1550-nm DFB diode laser, with a nominal output power of 40 mW. The driving current of the laser was adjusted in order to directly modulate the optical power at 5-Hz frequency and 13% modulation depth. The sensing fiber was composed by a 160 m-long standard single mode fiber (SMF), followed, in the first test, by a single 10-cm section of $Co^{2+}$-doped fiber, manufactured by CorActive HighTech, Inc. (ATN-FB series), whose absorption at 1550 nm was 50 dB/m.

A first test was carried out, aimed to demonstrate the capability of the system of carrying out simultaneous vibration and liquid detection measurements. The vibration was induced by wrapping a 5 m portion of the fiber around a PVC pipe, put in vibration using a magnetostrictive actuator bolted to the upper border of the pipe. The actuator was driven with a sinusoidal wave with a peak-to-peak current of 4 A, and a frequency of 100 Hz. For this test, the $Co^{2+}$-doped fiber section was surrounded by air. The C-OTDR measurements were carried out setting a pulse duration of 50 ns

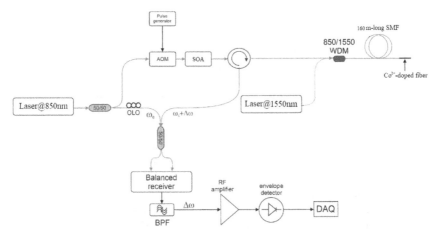

**Fig. 1** Experimental setup. AOM: acousto-optic modulator; SOA: semiconductor optical amplifier; WDM: wavelength division multiplexing; BPF: bandpass filter; DAQ: data acquisition card

(corresponding to a spatial resolution of 5 m), a pulse repetition period of 2 μs, and a number of averages equal to 1024. The combination of pulse repetition period and number of averages, resulted in an acquisition rate of ≈488 traces/s. The number of acquired traces was set to 512, corresponding to an overall acquisition time of ≈1 s. In Fig. 2a, we show the C-OTDR signals in the frequency domain, clearly revealing the region where the vibration was applied (z ≈ 62 m).

Furthermore, the 5-Hz signal induced by optical heating is observed at the $Co^{2+}$-doped fiber position (z ≈ 163 m). We also note that, some energy at 100-Hz is visible

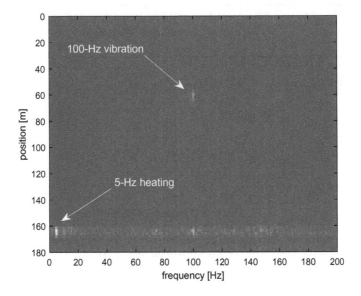

**Fig. 2** C-OTDR measurements along the entire fiber, displayed in the frequency domain

not only at the vibrated position, but also in correspondence with the $Co^{2+}$-doped fiber. This can be explained by the fact that, the signal received by the $Co^{2+}$-doped fiber is strongly amplified by the Fresnel reflection at the end of the sensing fiber, thus even a small phase modulation induced on it by the mechanical vibration, is captured at the detection stage.

The next test was carried out in order to demonstrate the capability of our sensor to discriminate the medium surrounding the $Co^{2+}$-doped specialty fibers. For this test, we have added another piece of 10-cm long $Co^{2+}$-doped fiber, at a distance of 10-m from the first one. The new measurements were carried out after increasing the number of averages up to 8192, resulting in an effective acquisition rate of $\approx$61 Hz. We have carried out 100 consecutive measurements, where, for each measurement the amplitude of the 5-Hz spectral component of the signal detected at each $Co^{2+}$-doped section was determined. The amplitude of the detected signal was calculated as the ratio between the 5-Hz spectral component of the C-OTDR signal at the monitored section, divided by the mean signal at the same section. The average and standard deviation of the detected signals are reported in Fig. 3, for three different surrounding media (air, water and acetone).

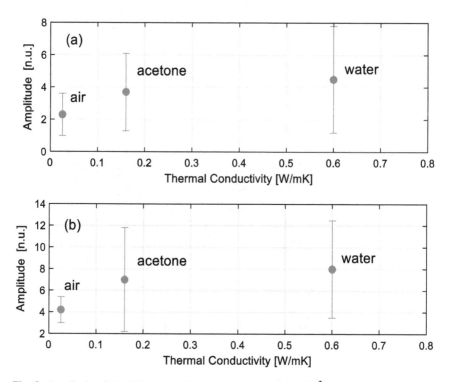

**Fig. 3** Amplitude of the 5-Hz spectral component at the **a** first $Co^{2+}$-doped fiber section and **b** second $Co^{2+}$-doped fiber section, as a function of the thermal conductivity of the surrounding medium. The error bars indicate the standard deviation over 100 consecutive measurements

The experimental results show that, while the single measurement is not able to distinguish the surrounding liquid, averaging over several measurements (100, in the present case) provides a signal increasing with its thermal conductivity, being the largest when the fiber is immersed into water. The physical reason is that, as the thermal conductivity increases, the surrounding medium is more efficient to dissipate away the heat conveyed by optical heating, resulting in a deeper temperature modulation. Figure 3 also reveals that the uncertainty of the measurements was quite large: this should be mostly attributed to the detection scheme used for the tests; in fact, C-OTDR sensors are strongly affected by fading and laser phase noise, which was quite large in our setup because of the relatively poor coherence of our laser source (3-MHz linewidth).

## 3 Conclusions

We have proposed and demonstrated a novel scheme for simultaneous vibration and liquid detection measurements. Vibrations are detected through a C-OTDR scheme operating at 850-nm wavelength, while liquid detection is done in selected positions based on active thermometry at 1550-nm wavelength. The advantage of the proposed method is that liquid detection can be performed in more points simultaneously, with limited additional cost with respect to a conventional C-OTDR scheme.

## References

1. Qin Z, Zhu T, Chen L, Bao X (2011) High sensitivity distributed vibration sensor based on polarization-maintaining configurations of phase-OTDR. IEEE Photonics Technol Lett 23:1091–1093
2. Mompó JJ, Martín-López S, González-Herráez M, Loayssa A (2018) Sidelobe apodization in optical pulse compression reflectometry for fiber optic distributed acoustic sensing. Opt Lett 43:1499–1502
3. Muanenda Y, Oton CJ, Faralli S, Di Pasquale F (2016) A cost-effective distributed acoustic sensor using a commercial off-the-shelf DFB laser and direct detection phase-OTDR. IEEE Photonics J 8:1–10
4. Hartog AH (2017) An Introduction to distributed optical fiber sensors. CRC Press.
5. Coscetta A, Catalano E, Cerri E, Cennamo N, Zeni L, Minardo A, A C-OTDR sensor for liquid detection based on optically heated Co2+-doped fibers. IEEE Sens J. https://doi.org/10.1109/JSEN.2020.2991224.
6. Davis MK, Digonnet MJF (2000) Measurements of thermal effects in fibers doped with cobalt and vanadium. J Lightwave Technol 18:161–165
7. Coscetta A, Catalano E, Cerri E, Zeni L, Minardo A (2020) Theoretical and experimental comparison of a distributed acoustic sensor at 850- and 1550-nm wavelength. Appl Opt 59:2219–2224

# Titanium Dioxide Doped Graphene for Ethanol Detection at Room Temperature

**Brigida Alfano, Maria Lucia Miglietta, Tiziana Polichetti, Ettore Massera, and Paola Delli Veneri**

**Abstract** The present work shows the synthesis of a graphene-based nanocomposite with titania nanoparticles through a simple one-step, microwave assisted method. The $GR/TiO_2$ nanocomposite was used for the fabrication of a chemiresistor device and characterized towards oxidizing and reducing gases under controlled environment. The results show how the addiction of titania nanoparticles to graphene nanosheets enables the sensing of ethanol at room temperature. A sensitivity curve to ethanol was recorded in the range 15–50 ppm. The interaction with water, the main interferent in ethanol sensing, was also investigated and the findings disclose a weaker interaction respect to ethanol, thus suggesting that steric factors can play a role in the sensing mechanism.

**Keywords** Graphene · Titanium dioxide · Graphene-metal oxide composite · Gas sensor · Ethanol

## 1 Introduction

Volatile Organic Compounds (VOCs) are recognized pollutants for their toxicity versus both the environment and human beings. As typical VOCs, alcohols are widely used in in medicine, especially during these days of Covid-19 outbreak, in chemical and in food industries, thus becoming a worldwide problem [1]. The pressing need for detection devices that could be easily deployable and integrated in portable devices has pushed the research towards the development of suitable gas sensors. In most cases, gas sensors based on metal-oxide semiconducting (MOS) materials are a suitable choice for alcohols detection; however, their high operating temperatures reduces the range of potential applications and devices in which they can be integrated. The research efforts in the last years were then devoted to the development of MOS devices capable to operate at room temperature. Research strategies in this field exploited mainly nanotechnologies, by means of novel features achieved by highly

B. Alfano · M. L. Miglietta (✉) · T. Polichetti · E. Massera · P. D. Veneri
ENEA, CR-Portici, P.le E. Fermi 1, 80055 Napoli, Italy
e-mail: mara.miglietta@enea.it

© The Author(s), under exclusive license to Springer Nature Switzerland AG 2021    107
G. Di Francia and C. Di Natale (eds.), *Sensors and Microsystems*,
Lecture Notes in Electrical Engineering 753,
https://doi.org/10.1007/978-3-030-69551-4_15

nanostructured metal oxides or introduction of materials having room-temperature chemical sensitivity [2–4].

As expected, the 2D graphene sheets, with its exceptional physico-chemical properties at room temperature including large surface area, high electrical conductivity and high chemical stability, is one of the most investigated material to this purpose [5–7]. Significant enhancements of the sensing performance have been actually attained by the use of hybrid materials made from metal oxides and graphene [4, 8]. In the main, sensitivity and selectivity are reported among the sensing features that benefit the most from the combination of the two materials [8, 9]. Moreover, the operating temperatures of sensors based on metal oxides are often decreased significantly upon addition of graphene [3].

The present work shows how the addiction of titania nanoparticles to graphene nanosheets enables the sensing of ethanol at room temperature. The hybrid material was prepared by a one-step, microwave assisted method and used for the fabrication of a chemiresistor device. The sensing layer was tested against oxidizing and reducing gases under controlled environment and the effect of humidity was also considered in relation to the ethanol sensing. A sensitivity curve to ethanol was recorded in the range 15–50 ppm.

## 2  Experimental

Pristine graphene nanosheets (GR) were prepared from natural graphite flakes (NGS Naturgraphit GmbH Winner Company) according to the method reported in [10]. The graphene nanocomposite with $TiO_2$ nanoparticles (Sigma-Aldrich, anatase, nominal purity 99.7%, primary particle size 25 nm) was prepared through a microwave assisted method described in ref [9] and illustrated in Fig. 1. The final GR/$TiO_2$ powder was dispersed in 2 ml of $H_2O$/IPA and this dispersion was drop-cast directly onto alumina transducers with interdigitated gold electrodes for device fabrication.

Morphological characterizations were conducted on films drop-cast on oxidized silicon and on silicon substrates. Raman spectra were taken by means of Renishaw InVia Reflex, with $\lambda = 514.5$ nm as excitation wavelength. Morphological characterizations and elemental microanalysis were performed by field-emission scanning electron microscopy coupled with energy dispersive X-ray spectrometer (FESEM/EDS LEO 1530–2 microscope, acceleration voltage 5 kV).

Chemiresistor devices were characterized under controlled environment in the Gas Sensor Test System developed by Kenosistec. Devices were biased at 1 V DC with a high stabilized generator TTI model QL355TP and the electrical current was measured and stored with Keithley Picoammeter 6485. Synthetic dry air was used as background gas and flown into the test chamber at 500 sccm, at constant temperature of 21 °C. The device were first stabilized in the background gas for 20 min, then exposed to the target gas for 10 min and finally restored under background gas for 20 min.

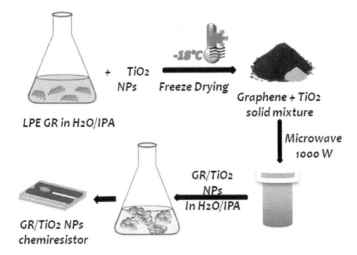

**Fig. 1** Preparation steps of GR/TiO$_2$ hybrid and of chemiresistors: dispersion of TiO$_2$ NPs in graphene solution, freeze drying and MW irradiation of the powder, re-dispersion in H$_2$O/IPA and drop-casting onto alumina substrate with interdigitated gold electrodes

## 3 Results and Discussion

In Fig. 2a the Raman spectrum of graphene functionalized through TiO$_2$ NPs (red line) is shown and compared with that of the pristine material (black line). The spectrum of GR evidences the characteristic features of few-layer graphene nanosheets, in which the 2D peak exhibits a fairly symmetrical shape, without the shoulder at about 2695 cm$^{-1}$ typical of the graphite 2D band. The downshift of both the G peak and the 2D band of GR/TiO$_2$ with respect to that of pristine GR indicates an effective interaction between TiO$_2$ NPs and graphene, as proof of its functionalization. In particular, the direction of peaks displacement indicates an n-type doping effect, which tends to neutralize or more precisely to reduce the p-type nature of GR. SEM image discloses a material with GR nanosheets mostly covered by TiO$_2$ nanoparticles

**Fig. 2** Raman spectrum of GR/TiO$_2$ (red line) compared to pristine graphene nanosheets spectrum (black line) (**a**) and SEM image (**b**) with related EDX analysis (**c**) of the GR/TiO$_2$ hybrid

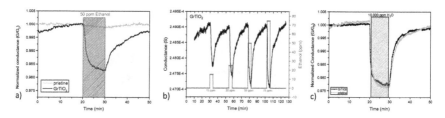

**Fig. 3** Dynamic responses: **a** single step exposure of GR/TiO$_2$ (red line) and of pristine graphene (black line) devices to 50 ppm of ethanol; **b** multiple step exposure of GR/TiO$_2$ chemiresistor to increasing concentrations of ethanol; **c** a single step exposure of GR/TiO$_2$ device to 10,000 ppm of water (i.e. 50% RH)

and bigger aggregates and an overall grainy surface. The Energy-dispersive X-ray (EDX) spectroscopy confirms the elemental composition of the nanocomposite.

The sensing behavior of GR/TiO$_2$ nanocomposite towards ethanol at room temperature was investigated and the results are summarized in Fig. 3. Under exposure to 50 ppm of ethanol for 10 min in dry environment the film conductance decreased, then slowly recovered in the next 20 min of synthetic air purging (Fig. 3a). This is not unexpected, since graphene has a well-known sensing ability at room temperature whereas the metal oxides usually requires higher temperatures to activate their sensing properties. Nonetheless, the sensitivity towards ethanol of this material was enabled by the presence of TiO$_2$ nanoparticles as can be inferred by the comparison with response of pristine graphene to ethanol (Fig. 3a).

When exposed to increasing concentrations of ethanol, the sensing response increased accordingly, as can be observed in Fig. 3b. The drift observed, is mainly caused by an incomplete recovery after each sensing cycle. Albeit not ideal, this behaviour can still be analyzed through an exponential fit during the detection and recovery phases in order to infer some of the basic features that characterize a sensing layer. In fact, under a constant perturbation, the chemiresistor undergoes an evolution between two stationary states and its change in conductance can be described by means of an exponential law. Hence, it is possible to estimate the characteristic times of the exponential growth and decrease of the electrical response (R–R$_0$/R$_0$), along with a sensitivity curve, both in the adsorption and in the desorption phases. These parameters are summarized in Table 1.

To further investigate the sensing behaviour of the GR/TiO$_2$ nanocomposite towards polar compounds, we tested the device versus water, which is moreover a ubiquitous interferent in any gas sensing scenario. The dynamic behaviour of the GR/TiO$_2$ sensing layer is reported in Fig. 4. As expected, an appreciable conductance

**Table 1** Estimated adsorption and desorption characteristic times of GR/TiO$_2$ for target gases

| Analyte | $t_{ads}$ (s) | $S_{ads}$ (%/ppm) | $t_{des}$ (s) | $S_{des}$ (%/ppm) |
|---------|---------------|-------------------|---------------|-------------------|
| EtOH | 80 | $4 \times 10^{-2}$ | 250 | $2 \times 10^{-2}$ |
| H$_2$O | 44 | $2 \times 10^{-4}$ | 129 | $2 \times 10^{-4}$ |

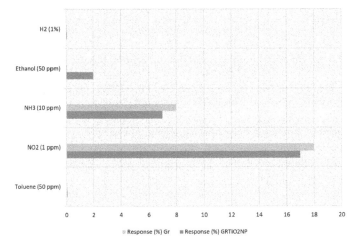

**Fig. 4** Selectivity of GR/TiO$_2$ respect to GR-based device to different gases at room temperature

decrease is observed upon 10 min exposure to 10,000 ppm of water (corresponding to a relative humidity of 50%). The recovery, in this case, appears much faster. Indeed, the estimated adsorption and desorption times for water are halved with respect to the ethanol sensing (Table 1).

Noticeably, the sensitivity towards these two analytes is rather different, in fact, the sensitivity to ethanol is 200-fold greater than that towards water.

The effect of graphene decoration with the TiO$_2$ nanoparticles is then reflected mainly in the material specificity to different gases (Fig. 4). It can be observed that the basic sensing features of graphene are preserved while it is worth noting the arising of a sensitivity towards ethanol.

The analysis reported in Table 1 highlights high desorption times for ethanol, comparable to those previously observed for NO$_2$, thus suggesting comparable inter-action energies for both compounds, albeit of different natures [11]. Ethanol has in fact only a minimal interaction with graphene layers, as can be inferred by the very low response of pure GR-based device.

The sensitivity to ethanol can be regarded then as an effect of the surface character-istics and morphology of the hybrid. The microwave irradiation during the synthesis is likely to have gained the hydrophilic nature of TiO$_2$ particles, by producing vacan-cies or Ti ions, which improve the chemical affinity towards polar compounds and hydroxyl-bearing substances in particular [12]. The observed sensitivity to ethanol can then arise from this enhanced interaction. Afterward, the conductivity change can be ascribed mainly to a swelling effect, with the ethanol adsorbed onto the TiO$_2$ phase and swelling the surface with a disruption of the conductive pathways, finally leading to the observed conductance decrease. In this view, water, with its smaller steric hindrance has a far less effective swelling effect, as reflected in the lower sensitivity.

# 4 Conclusion

The results presented herein show that the sensing layers based on the GR/TiO$_2$ hybrid present the synergistic effects of the two components. The basic sensing features of the graphene nanosheets are preserved whereas some new specific features are enabled by the presence TiO$_2$ NPs. The sensitivity towards low concentration of ethanol at room temperature is indeed a characteristic signature of the hybrid material. The interaction with water, the main interferent in ethanol sensing, appears much weaker respect to ethanol, thus suggesting that steric factors can play a role in the sensing mechanism. Further investigations are needed to clarify the sensing mechanism of these polar compounds.

# References

1. Willey JD, Avery GB, Felix JD, Kieber RJ, Mead RN, Shimizu MS (2019) Clim Atmos Sci 2:1–5
2. Qin YX, Fan GT, Liu KX, Hu M (2014) Sens Actuators B 190:141
3. Basu S (2020) Multilayer thin films-versatile applications for materials engineering. IntechOpen
4. Sun D, Luo Y, Debliquy M, Zhang C (2018) Beilstein J Nanotechnol 9:2832–2844
5. Schedin F, Geim AK, Morozov SV, Hill EW, Blake P, Katsnelson MI, Novoselov KS (2007) Nat Mater 6:652
6. Novoselov KS, Geim AK, Morozov SV, Jiang D, Zhang Y, Dubonos SV, Grigorieva IV, Firsov AA (2004) Science 306:666
7. Tian W, Liu X, Yu W (2018) Appl Sci 8:1118
8. Chatterjee SG, Chatterjee S, Ray AK, Chakraborty AK (2015) Sens Actuators B 221:1170–1181
9. Alfano B, Miglietta ML, Polichetti T, Massera E, Bruno A, Di Francia G, Delli Veneri P (2019) IEEE Sens J 19:8751–8757
10. Fedi F, Miglietta ML, Polichetti T, Ricciardella F, Massera E, Ninno D, Di Francia G (2015) Mater Res Express 2(3):035601
11. Alfano B, Polichetti T, Mauriello M, Miglietta ML, Ricciardella F, Massera E, Di Francia G (2016) Sens Actuators B: Chem 222:1032–1042
12. Kabongo GL, Nyongombe G, Ozoemena K, Dhlamini S (2018) J Nanosci Curr Res 3. 2572-0813

# Gold Coated Nanoparticles Functionalized by Photochemical Immobilization Technique for Immunosensing

**B. Della Ventura, R. Campanile, M. Cimafonte, V. Elia, A. Forente, A. Minopoli, E. Scardapane, C. Schiattarella, and R. Velotta**

**Abstract**  Gold nanoparticles (AuNP)s play a fundamental role in biosensing in view of their various applications, each of them exploiting one or more properties of such a system. In any case, to make AuNPs effective as biosensing tools, it is necessary to provide them with high specificity by means of processes able to achieve an efficient gold surface functionalization with antibodies. This issue can be easily addressed by the so-called Photochemical Immobilization Technique (PIT). Herein, we report PIT-functionalized AuNPs applied as ballast for improving the sensitivity of quartz-crystal microbalances, but also as optical transducers (colorimetric biosensor) in which the key role is played by the plasmonic properties of nanostructured gold. Eventually, the role of gold is highlighted in its combination with magnetic nanoparticles, a physical system in which the magnetic behaviour of the core is joined to the optical properties of the surface.

**Keywords**  Biosensors · Nanoparticles · Core-shell nanoparticles · Photochemical immobilization technique · Colorimetric · Localised surface plasmon resonance

## 1  Introduction

Gold nanoparticles (AuNP)s are being largely used in biosensing thanks to their inertness and plasmonic properties, which make them highly exploitable in manifold ways [1]. One of the key steps to be taken into account when using AuNPs is their functionalization, whereby they specifically interact with the analyte to be detected. Although molecularly imprinted polymers (MIP)s are attracting wide interest in view of their stability [2], "natural" bioreceptors, and particularly antibodies (Ab)s, still represent the most effective choice for their unique capability of selectively

---

B. Della Ventura · R. Campanile · M. Cimafonte · V. Elia · A. Forente · A. Minopoli ·
E. Scardapane · C. Schiattarella · R. Velotta (✉)
Dipartimento di Fisica Ettore Pancini, Università degli Studi di Napoli Federico II, Via Cintia, 26, 80126 Naples, Italy
e-mail: rvelotta@unina.it

G. Di Francia and C. Di Natale (eds.), *Sensors and Microsystems*,
Lecture Notes in Electrical Engineering 753,
https://doi.org/10.1007/978-3-030-69551-4_16

interacting with a single analyte (antigen) thereby providing unparalleled specificity [3].

A typical Ab used in biosensing is the immunoglobulin G (IgG), a protein with a molecular weight of approximately 150 kDa, tens nanometers in size, and a characteristic Y-like shape (Fig. 1a). For biosensing purposes, the most important domain is the so-called Fragment antigen binding (Fab) (depicted in red in Fig. 1a) that constitute the antigen-binding sites of the Ab providing its characteristic high specificity [4]. Of course, when dealing with the design of a biosensing platform that exploits Abs, a spatial orientation that leaves at least one of the two Fab exposed to the solvent is mandatory; alternatively, both Fabs anchored onto the biosensor surface would be highly detrimental for the device sensitivity since no recognition can take place. Moreover, an effective functionalization would require a high Abs surface density so to fully exploit the interacting surface of the biosensor. Currently, even by using complex and time-consuming chemical functionalization procedures, satisfactory results are not warranted [5–7].

In an effort towards the achievement of a simple and effective gold surface functionalization with Abs, we set up the Photochemical Immobilization Technique (PIT), a two-steps procedure that includes the activation of whole IgGs in solution by high intensity UV lamp and their subsequent transfer onto the biosensor surface for straightforward Ab surface binding (Fig. 1a) [8, 9]. PIT is based on the selective photoreduction of the disulphide bridge in some of the cysteine–cysteine/tryptophan (Cys-Cys/Trp) triads [10], which are a typical structural feature of the IgG [11]. The UV excitation of the Trp residue leads to the generation of solvated electrons, which are captured by the nearby disulphide bridge resulting in its destabilization and subsequent breakage of the cysteine-cysteine bond. The as-produced free thiol groups interact with proximal gold surface giving rise to an oriented covalent IgG immobilization (Fig. 1b). This strategy has been shown to ultimately enhance the antigen detection efficiency of immunosensors based on quartz-crystal microbalances [12] as well as on screen-printed electrodes for electrochemical sensing [13]. Herein, we review our recent results that show how PIT can be successfully applied to colloidal AuNPs to provide either a "ballasting" tool for mechanical platforms or colorimetric transducers.

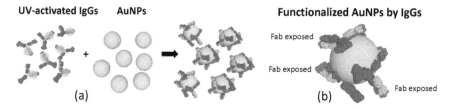

**Fig. 1** Schematic representation of the functionalization process. **a** Antibodies (IgG)s are activated by UV irradiation and mixed to a solution containing AuNPs. After short incubation AuNPs are functionalized with Abs. **b** Magnification of a single AuNP highlighting the heavy chain (green), light chain (blue) and Fab (red) domain, the latter being exposed to the solvent for all the three Abs

## 2 AuNPs Synthesis and Functionalization

### 2.1 AuNPs Synthesis

Spherical gold nanoparticles were synthesized by modifying an existing protocol [14]. In short, we added sodium citrate (20 mg/mL) to a previously prepared gold salt solution (10 mg/mL of $HAuCl_4 \cdot 3H_2O$) to achieve particles nucleation. After that, particles growth was induced with continued stirring and boiling until the colour of the suspension turned dark red. The reached optical density was approximately 1.0 at 530 nm, corresponding to a concentration of approximately $10^{10}$ particles/mL [15].

### 2.2 Synthesis of Core-Shell Magnetic NPs (Au@MNP)

The synthesis consisted in mixing 75 μL (5 mg/mL) of $Fe_3O_4$ and 1 mL (10 mg/mL) of sodium citrate dihydrate in 15 mL of milliQ water at 120 °C under a slow stirring. When the solution boiled, 4 additions of 50 μL (10 mg/mL) of $HAuCl_4$ every 5 min were performed. The first step produced the gold seeds that anchored to the magnetic core, while the second step induced the growth of the gold layer. The formation of the Au shell was signalled by a color change of the solution from brownish to burgundy.

### 2.3 Functionalization of Nanoparticles by PIT

1 mL of AuNPs (or core-shell NPs) was centrifuged at 3000 g for 10 min. The resulting supernatant was discarded and the AuNP pellet was resuspended in ultra-pure water. Simultaneously, a solution of 50 μg/mL IgG was prepared and irradiated for 30 s with the UV lamp (Trylight®) at 254 nm (Fig. 2a). The power of each lamp is 6 W so that the intensity on the quartz cuvette is approximately 0.3 $W/cm^2$.

A volume of 20 μL of the irradiated IgG solution was added to 1 mL of AuNPs, and the resulting mixture was incubated for 3 min. The amount of IgG in the mixture was progressively increased by further adding 4 μL of the irradiated Ab solution. The UV-vis absorption spectra showed a red shift of the maximum absorption wavelength as a result of the formation of the protein corona (Fig. 2b). Only few spikes were necessary to stabilize the red shift; thus, a total number of five additions, corresponding to 1 μg/mL IgG concentration, demonstrated to ensure the maximum coverage of the AuNP surfaces. Unbound Abs were removed from the mixture by centrifugation at 3000 g for 10 min (low temperature) and supernatant discarding. The IgG-functionalized AuNP pellet was resuspended in ultrapure water. Finally, 1 mg/mL of bovine serum albumin (BSA) was added to the functionalized colloid to block the AuNP surface from nonspecific adsorption.

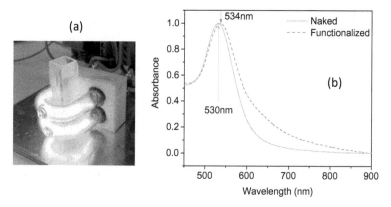

**Fig. 2** **a** TryLight® lamp used for PIT. The intensity in the cuvette is approximately 0.3 W/cm².
**b** Absorption spectrum of a colloidal solution of AuNPs before (red continuous line) and after
functionalization (dashed blue line) (color figure online)

# 3   Results

A direct application of AuNPs exploits the high density of gold in a detection scheme
depicted in Fig. 3, which uses a quartz-crystal microbalance as mechanical transducer
for small molecules (<300 Da) detection. The target (violet circle) is detected thanks
to the highly specific ballast provided by functionalized AuNPs in the sandwich
configuration Ab-analyte-AuNP. In such a way, parathion in drinkable water could

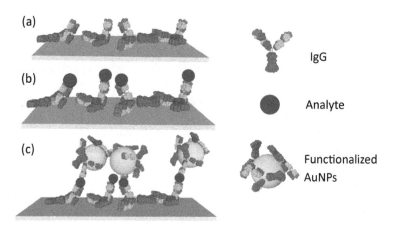

**Fig. 3** **a** The gold surface is functionalized with PIT. The IgGs are close packed (high surface
density) and expose one Fab to the solvent (upright orientation) so to provide an ideal functionalization. **b** The Abs recognize the analytes with small or even undetectable signal if the latter are
small. **c** Strong signal is detected when AuNP bind the analyte leading to a highly specific detection
(color figure online)

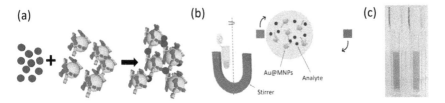

**Fig. 4 a** The presence of the analyte (violet circle) induces aggregation of functionalized Au@MNPs. **b** A rotating magnet improves the mobility of magnetic nanoparticles thereby increasing the yield of aggregate occurrence. **c** A colloidal solution of Au@MNPs (left) changes its color when 2 μL of IgG solution (250 ng/mL) is added (right) (color figure online)

be detected with a limit of detection lower than 1 ppb [16], with an improvement of more than one order of magnitude in comparison with 15 ppb previously measured with a simple sandwich Ab-analyte-Ab [17].

The highly effective gold surface functionalization offered by PIT shows its potential when AuNPs are exploited for their plasmonic properties. The detection scheme is depicted in Fig. 4a and shows AuNP aggregation induced by the presence of the analyte that acts as a linker. A solution of 1 mL of functionalized AuNPs was mixed to a smaller volume of solution containing the analyte (typically 100–200 μL). This is helpful in all the situations, as it is the case for tap or drinking water, in which the analyte solution contains salts, which may induce (non-specific) aggregation. With this approach, we were able to realize a colorimetric biosensor for 17β-estradiol in tap water with a limit of detection of only few pg/mL [18].

Additional properties can be provided to the user when gold is exploited as coating of a magnetic nanoparticle (MNP), so to realize a core-shell nanoparticle (Au@MNP). Indeed, these nanoparticles can be "steered" by a magnetic field while retaining all the plasmonic properties. In fact, the gold surface of Au@MNPs lends themselves to be easily functionalized with antibodies by PIT, thereby becoming promising "analyte catchers" at the nanoscale with inherent application to biosensing devices. To test the occurrence of both optical and magnetic properties, we used Au@MNPs (size ≈ 50 nm with an inner diameter of 30 nm) to reduce the limit of detection of a colorimetric immunosensor previously developed [19].

The rationale of our approach relies on the higher nanoparticle mobility induced by a microstirrer realized in a microtube placed off-axis with respect to a rotating magnet (see Fig. 4b). The rotating magnetic field acts as external force, which pushes the otherwise slow nanoparticles so to increase the collision rate among Au@MNPs and analytes. This, in turn, is directly related to the efficiency with which the aggregates are formed and, hence, to the limit of detection. Such an improvement is demonstrated by the color change that takes place when IgG solution at 250 ng/mL is mixed to a colloidal solution of Au@MNPs (Fig. 4c). Besides, no effect could be observed when simple AuNPs of the same size were used (data not shown). It is worth noticing that the stirring realized by our approach takes place in a volume as small as 50 μL.

# 4 Conclusion

Surface functionalization is a key step in realizing a biosensor, which we address by adopting the Photochemical Immobilization Technique (PIT), a simple and effective procedure able to tether antibodies upright on a gold surface. In this manuscript, we report PIT application to AuNPs, showing that the resulting functionalized nanostructures are effective in binding antigens; thus, they can be successfully used in a large variety of biosensing applications such as ballasting tool for quartz crystal microbalance or in colorimetric assays. Eventually, we show that gold-coated core-shell magnetic NPs can also be functionalized by PIT, allowing mixing in volumes as small as 50 µL (microstirrer) under application of a proper external field. Such a procedure can improve the limit of detection in colorimetric biosensors.

# References

1. Malekzad H, Sahandi Zangabad P, Mirshekari H, Karimi M, Hamblin MR (2017) Nanotechnol Rev 6:301
2. Uzun L, Turner APF (2016) Biosens Bioelectron 76:131
3. Vashist SK, Luong JHT (2018) Handb. Immunoass. Technol. Elsevier, Weinheim, pp 1–18
4. Conroy PJ, Hearty S, Leonard P, O'Kennedy RJ (2009) Semin Cell Dev Biol 20:10
5. Jung Y, Jeong JY, Chung BH (2008) Analyst 133:697
6. Vashist SK, Dixit CK, MacCraith BD, O'Kennedy R (2011) Analyst 136:4431
7. Welch NG, Scoble JA, Muir BW, Pigram PJ (2017) Biointerphases 12:02D301
8. Della Ventura B, Schiavo L, Altucci C, Esposito R, Velotta R (2011) Biomed Opt Express 2:3223
9. Della Ventura B, Banchelli M, Funari R, Illiano A, De Angelis M, Taroni P, Amoresano A, Matteini P, Velotta R (2019) Analyst 144:6871
10. Neves-Petersen MT, Gryczynski Z, Lakowicz J, Fojan P, Pedersen S, Petersen E, Bjørn Petersen S (2002) Protein Sci 11:588
11. Ioerger TR, Du C, Linthicum DS (1999) Mol Immunol 36:373
12. Fulgione A, Cimafonte M, Della Ventura B, Iannaccone M, Ambrosino C, Capuano F, Proroga YTR, Velotta R, Capparelli R (2018) Sci Rep 8:16137
13. Cimafonte M, Fulgione A, Gaglione R, Papaianni M, Capparelli R, Arciello A, Bolletti Censi S, Borriello G, Velotta R, Della Ventura B (2020) Sensors 20:274
14. Pollitt MJ, Buckton G, Piper R, Brocchini S (2015) RSC Adv 5:24521
15. Haiss W, Thanh NTK, Aveyard J, Fernig DG (2007) Anal Chem 79:4215
16. Della Ventura B, Iannaccone M, Funari R, Pica Ciamarra M, Altucci C, Capparelli R, Roperto S, Velotta R (2017) PLoS One 12:e0171754
17. Funari R, Della Ventura B, Carrieri R, Morra L, Lahoz E, Gesuele F, Altucci C, Velotta R (2015) Biosens Bioelectron 67:224
18. Minopoli A, Sakač N, Lenyk B, Campanile R, Mayer D, Offenhäusser A, Velotta R, Della Ventura B (2020) Sens Actuators B Chem 308:127699
19. Iarossi M, Schiattarella C, Rea I, De Stefano L, Fittipaldi R, Vecchione A, Velotta R, Della Ventura B (2018) ACS Omega 3:3805

# Concept of MEMS Vibrating Membrane as Particulate Matter (PM) Sensor

**Francesco Foncellino, Luigi Barretta, and Ettore Massera**

**Abstract** This work shows experimental results obtained in testing a thin film piezo actuated silicon membrane in a controlled environment. The main aim of this work was to select humidity concentration as one of the most influencing factors on vibrating characteristic of the circular clamped membrane. Thanks to fine control of humidity and the parallel feedback due to calibrated sensors we established a precise correlation between humidity and resonant shift.

**Keywords** Gas sensors · Calibration · Embedded system

## 1 Introduction

The problem of environmental particulate matter (PM) has reached high levels of attention above all for problems relating to the respiratory tract. There are several macro devices on the market that provide PM detection while there is still plenty of room for the development of microscopic detectors. Currently many research groups are using a low-cost PM sensor that can be integrated with portable technologies [1]. Resonant sensors based on microelectromechanical systems (MEMS) have attracted attention to implement a small compact sensing system thanks to their high mass sensitivity and simple processing [2]. They can be realized according various available technologies and type of driving elements integrated into cantilevers, including mechanical piezoceramic, electrostatic, piezoelectric and electrothermal components to excite them in resonance.

F. Foncellino (✉)
STMicroelectronics, via Remo de Feo, 80022 Arzano (NA), Italy
e-mail: francesco.foncellino@st.com

L. Barretta
Dipartimento di Fisica, Università degli Studi Napoli "Federico II, Naples, Italy

E. Massera
ENEA, CR-Portici, P.le E. Fermi 1, 80055 Naples, Italy

© The Author(s), under exclusive license to Springer Nature Switzerland AG 2021    119
G. Di Francia and C. Di Natale (eds.), *Sensors and Microsystems*,
Lecture Notes in Electrical Engineering 753,
https://doi.org/10.1007/978-3-030-69551-4_17

As a global semiconductor player, STMicroelctronics is expanding its product portfolio and is committed to the development of piezo actuated devices.

The population of these devices is growing rapidly and the pMUT (micromachined piezoelectric ultrasound transducer) occupies a special place particularly suitable for applications such as medical imaging, gesture sensors, fingerprint and ultrasound sensors, body composition sensors, object detection etc.

A pMUT is a thin piezoelectric film stacked between two electrodes on a passive membrane that can vibrate. It is able to obtain transmission and acoustic reception based on its vibrating diaphragm operated by a piezoelectric material.

Apart the focused goal of a pMUT we can admit that its structure can be useful for a different aim: the appropriately sized vibrating membrane can be used as a PM sensor using its resonant properties as a function of the particles that are deposited on the surface.

Taking advantage of their vibrational properties, PMUT can be used as a particulate sensor and once optimized the design is a good candidate for the development of portable technology. Its promising characteristics definitely represent an innovative approach to PM sensing technology commercially available on the market as for example quartz microbalance. The resonant MEMS sensors offer: excellent sensitivity, miniaturized device size, quasi-digital frequency output and the ability of being integrated into an on-chip sensor array.

The first step for the realization of an innovative device for the identification and evaluation of PM based on PMUT is the characterization of the membrane in a controlled environment.

In this work we present a first approach for the characterization in controlled environment of the ST Microelectronics's Pmut matrix to understand the capability of this device as PM sensor. We begin with the study of the impedance spectrograms and how they change with the moisture content.

## 2 Experimental and Results

### 2.1 Experimental Setup

Prototypes of thin film circular clamped piezoactuated membrane realized by STMicroelectronics is assembled onto a dedicated PCB that is inserted in the testing chamber (Fig. 1). A Vector Network Analyzer (E5061B Keysight) is connected to PCB by means of external cables. A PM sensor (NOVASENSE SDS011) is put in the chamber with the aim to monitor the internal particulate concentration.

Temperature and humidity are recorded with industrial sensors (LSI Pt100) humidity sensor.

The chamber is kept at 21.5 °C and is flushed with a mixture of gas that provides the relative humidity (RH) according the planned set point.

**Fig. 1** Photo of the 15Lt "Large Volume Test Chamber" (LVTC) with the **a** PCB and **b** NOVASENSE SDS011

The Network Analyzer provides the impedance spectrum at each steady state point. The membrane is continuously excited with a sinusoidal sweeping signal and its impedance its monitored. When the humidity set point is reached, we record the impedance and extract all the first resonance parameters with the main focus on the maximum of the phase.

The experimental procedure is characterized by a first RH measurement that was of 14.6% then we reached the 9.1% set point.

Data acquisition is carried out varying humidity levels in steps of 5% in forward and backward direction from 9.1 to 80%. Once reached the 80% we started to lowering the RH to the initial values always proceeding in steps of 5% variation in humidity.

Anyway due to measurement repeatability we focused and report only data about the backward direction where we focused our analysis on maximum of impedance phase (Fig. 2).

## 2.2 Results

Once the measures were obtained, the data were analyzed. We analyzed the behavior of the maximum of the phase during the backward phase (Fig. 3).

As you can see from the picture the resonance shift is well fitted by a polynomial law as already reported [3].

The shift in frequency, due to the relative humidity, will be taken into account as a correction when the membranes will be approved for the measurements of particulate, whenever the shift amplitude due to the second phenomenon would be comparable or a little bit bigger to the shifts due to the RH variations. The relative humidity will be measured with another device (MEMS), likely a commercial device, and so it will be correct via software or simply thanks to electronics.

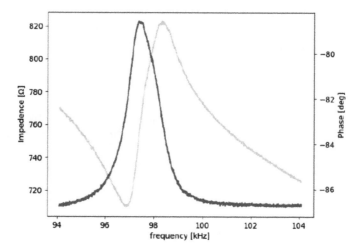

**Fig. 2** Plot of the impedance and phase spectrum of the analized PMUT in the resonance region

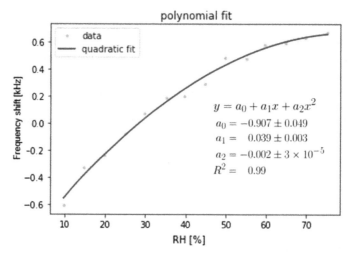

**Fig. 3** Quadratic FIT of the Frequency shift of the measured phase maximum due to relative humidity change

## 3 Conclusions

The measurements carried out showed that the resonance of the membrane is greatly affected by the humidity. In addition, we were able to identify the quadratic relation of influence and evaluate the maximum phase shift that is about 1.3 kHz.

A more detailed analysis will be led considering the foretold forward phase and then a complete analysis will be done comparing forward and backward directions.

# References

1. Zampetti E et al (2017) Exploitation of an integrated microheater on QCM sensor in particulate matter measurements. Sens Actuators A 264:205–211
2. Ismail AK (2006) The principle of a MEMS circular diaphragm mass sensor. J Micromech Microeng 16:1487–1493
3. Sun et al (2019) Self-powered multifunctional monitoring system using hybrid integrated triboelectric nanogenerators and piezoelectric microsensors. Nano Energy 58:612–623

# Assessing the Impact of Haze on Solar Energy Potential Using Long Term PM$_{2.5}$ Concentration and Solar Insolation Filed Data in Naples, Italy

M. Nocerino, G. Fattoruso, G. Sorrentino, V. Manna, S. De Vito, M. Fabbricino, and G. Di Francia

**Abstract** Atmospheric fine particulate pollutant affects seriously the human health but also the passage of light through the lower atmosphere, reducing the solar radiation reaching the ground as well as the PV panels. In this study, the solar insolation reduction due to air pollution has been investigated in the city of Naples (Italy). Analyzing local long term field data, we have obtained that the solar insolation reduction is exponentially correlated to the PM2.5 concentration. By using the derived empirical relation, for Naples it was estimated that the solar insolation was reduced around 5% or 66.20 kWh/m$^{-2}$ within one-year period (May 2018–May 2019), due to air fine particulate pollution. This study provides the theoretical basis to design successful solar PV systems to be mounted on building rooftops or in other suitable sites, taking into account also the local air pollution condition.

**Keywords** Solar radiation · Solar photovoltaic potential · Atmosphere particulate pollution

## 1 Introduction

Air pollution is known to be a serious problem mainly for its effects on human health. Recently, however, the air pollution effects in other fields such as the photovoltaic energy generation have been investigated and, notably, the relation between the reduction of the solar insolation reaching the PV systems and fine particular matter (PM$_{2.5}$) concentration in the air. Air pollution typically includes PM$_{2.5}$ which suspended in the atmosphere can reduce the solar radiation intensity reaching the ground. At this regard, Peters et al. [1], correlating measured particulate concentrations and solar insolation in Delhi over a long period of 19 months, estimated that

G. Fattoruso · S. De Vito · G. Di Francia
ENEA, Research Centre Portici, P.le E. Fermi 1, 80055 Naples, Italy
e-mail: grazia.fattoruso@enea.it

M. Nocerino (✉) · G. Sorrentino · V. Manna · M. Fabbricino
University of Naples Federico II, Via Claudio, 21, Naples, Italy

total sunlight reaching the ground during one year was reduced by more than one ninth, due to air pollution.

Wang et al. [3], testing a distributed photovoltaic system on a building roof in Shanghai, showed that the higher the $PM_{2.5}$ concentration, the lower the power generation capacity of the PV module was. Then, the solar radiation received on the surface of the PV module resulted exponentially related to the atmospheric $PM_{2.5}$ concentration.

It is known that the cities in the Eastern Mediterranean and Southeast Asia are often affected by major haze events reaching $PM_{2.5}$ concentrations up to 375 $\mu g\, m^{-3}$ [1, 2]. However, most of cities around the world suffers the air pollution although at several seriousness levels.

In this study, we intend at addressing the impact of $PM_{2.5}$ particles on solar insolation levels in the city of Naples (South Italy), whose air pollution levels exceed often the threshold values.

By 2030, EU countries will have to increase the use of renewables, including the solar PV systems, for realizing the 32% target of renewable energy production, according to the EU's re-cast renewable energy directive [4]. In this regard, EU orientation is to address the use of suitable buildings' surfaces rooftops and facades for distributed solar PV systems deployments.

Recently, Bodis et al. [5] have estimated that the EU cities' building rooftops could potentially produce solar PV electricity, annually covering the 25% of the current electricity. In particular, Italy could potentially cover more than 30% of its electricity consumption by developing rooftop PV systems at its most advantageous rooftops. In view of exploitation of this potential, the estimation of the reduction of solar radiation due to $PM_{2.5}$ particles in the air could be fundamental, making the difference between a solar PV installation meets the expected output and one that fails. In this study, at this scope, an empirical relation between $PM_{2.5}$ concentration and reduction in insolation has been derived by analyzing one-hour $PM_{2.5}$ and insolation measurements from a monitored location in Naples, Italy.

## 2   Correlating $PM_{2.5}$ Concentration and Reduction in Solar Radiation

The collected data used in our analysis were recorded over 19 months between January 2018 and July 2019. Insolation data were measured by a pyranometer with a frequency of one measurement every hour. This instrument consists in a thermopile sensor coated with an opaque black paint providing a flat spectral response for the full wave length range. It measures the global solar radiation on a plane/level surface as sum of direct solar radiation and diffuse sky one.

Fine particulate data were recorded by an air quality monitor as SWAM 5A DUAL CHANNEL with one-hour frequency.

**Table 1** PM2.5 concentration ranges and corresponding AQI color code

| Concentration range ($\mu$g m$^{-3}$) | Color code | Levels of health concern |
|---|---|---|
| 0-20 | Green 1 | Extremely Good |
| 21-30 | Green 2 | Very Good |
| 31-40 | Green 3 | Moderate Good |
| 41-50 | Green 4 | Good |
| 51-100 | Yellow | Moderate |
| 101-150 | Orange | Moderate unhealthy |
| 151-200 | Red | Unhealthy |
| 201-300 | Purple | Very unhealthy |
| Above 300 | Maroon | Hazardous |

Both instruments are located at the air quality gauge station, located in Naples, managed by ARPAC (Italian Agency of Environmental Protection). This station is identified by the EU code IT1493A.

The approach, used for relating $PM_{2.5}$ concentration and insolation data, consists in three main steps: (1) normalizing insolation data, (2) clustering $PM_{2.5}$ concentration data, (3) deriving the correlation curve.

At first, $PM_{2.5}$ concentration levels were classified according the color coding of Air Quality Index (AQI). We observed that most of recorded concentration levels fell in the first AQI class ranging among 0–50 $\mu$g m$^{-3}$. For making the analysis consistent, we split the first class in sub-classes as showed in Table 1, transferring more properly the recorded data in AQI levels.

Hence, insolation data were sorted in bins corresponding to the defined different $PM_{2.5}$ concentration levels. Then, humidity and clear sky filters were used for identified data representative of clear sky conditions.

It is to be noted that, over the course of a year, the insolation varies via the zenith angle and the eccentricity of the Earth's orbit around the sun as well as due to seasonal variations in atmospheric conditions. Since insolation data covered a longer period than one year, these variations were to be considered in the correlation analysis. At this scope, insolation data were analyzed for each month separately and normalized by taking the 90 percentile value of all insolation measurements at noon for all months.

Figure 1 shows the results of the normalization for the most representative months of June and November.

For identifying the conditions representative of a clear sky, the 80 percentile filter of the datasets for each hour and pollution level was calculated.

This data was used for deriving the correlation curve between $PM_{2.5}$ concentration levels and reduction of insolation, considering the relative reduction for each hour. For generating this functional relation, only datasets containing more than seven data points were considered.

The data analyzed were fitted by an exponential decay curve with a value $R^2$ of 55%.

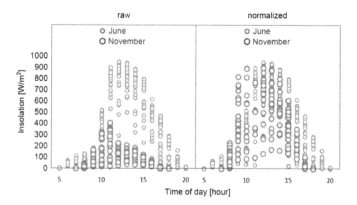

**Fig. 1** Graphs of insolation data. On the left, the raw data and on the right the data normalized, corresponding at June and November months

Exponential decay was expected according to Lambert–Beer's law. The fitted exponential decay is shown as a black line in Fig. 2.

The fitted curve is described by the following equation:

$$\frac{I(PM_{2.5})}{I_o} = \exp\left(\frac{-PM_{2.5}}{250}\right) \tag{1}$$

where $I_0$ is the isolation at 0 $\mu$g m$^{-3}$ and $I$ is the insolation affected by PM2.5 concentration.

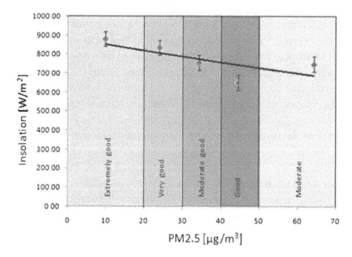

**Fig. 2** Exponential variation of solar radiation with PM$_{2.5}$ concentration

As simplification, this result is obtained assuming consistent, over the entire period considered, the composition and size distribution of air pollution as well as the optical behavior of the aerosol.

## 3 Estimating PM$_{2.5}$ Related Reduction in Solar Insolation

Equation (1) was used to estimate over one year how much light is lost due to PM$_{2.5}$ related air pollution in Naples.

Figure 3 shows the measured insolation and the insolation estimated by means of Eq. (1), considering PM$_{2.5}$ concentration of 0 $\mu$g m$^{-3}$ ($I_0$), over one year from May 2018 to May 2019. Integrating this insolation data, we calculated the annual insolation and the projected insolation at 0 $\mu$g m$^{-3}$ PM$_{2.5}$, as shown in Table 2.

So, we found that the amount of insolation for Naples was reduced by 5% or 66.20 kWh/m$^{-2}$ of the annual solar energy reaching the ground.

From this result, we can derive that the air fine particulate pollution at concentration levels of 0–100 $\mu$g m$^{-3}$ can affect the solar energy potential of viable rooftops for PV installations. Moreover, using this empirical relation, the insolation loss due to air quality can be directly evaluated, making more reliable the solar PV potential estimations [6].

**Fig. 3** Measured insolation and estimated insolation without air pollution due to PM$_{2.5}$ over one year

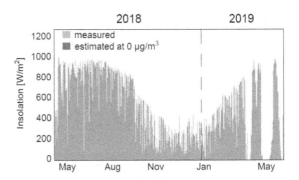

**Table 2** PM2.5 effects on solar insolation in Naples

| Annual radiant exposure | |
| --- | --- |
| Projection 0 $\mu$g m$^{-3}$ | 1368.41 (kWh m$^{-2}$) |
| Measured data | 1302.21 (kWh m$^{-2}$) |
| Difference | 66.20 (kWh m$^{-2}$) |
| Loss in percentage | 4.84% |

# 4 Conclusions

Atmospheric particulate pollutant can affect the transition of light through the lower atmosphere, reducing the solar radiation reaching a PV panel. So, the variation of solar radiation intensity with $PM_{2.5}$ concentration, mainly in the cities, is investigated in order to make more reliable PV power estimations.

In this work, we have analyzed the correlation between $PM_{2.5}$ concentration and the loss in isolation, using long term field data from a monitored location in Naples. As result, we obtained that the amount of solar radiation is exponentially correlated to the $PM_{2.5}$ concentration.

Using the formulated functional relation, we calculated the amount of insolation Naples would have received without the air particulate pollution within one-year period. In particular, we estimated that the insolation due to $PM_{2.5}$ pollution was reduced around 5%, from 1368 to 1302 kW m$^{-2}$.

The derived empirical relation represents itself a result. It enables a way to predict the influence of air fine particulate pollution on the solar energy potential in different PV suitable sites.

In conclusion, this study provides the theoretical basis to design solar PV systems to be mounted on a building rooftop as well as in other suitable sites, taking also into account the local air pollution condition.

# References

1. Peters IM, Karthik S, Liu H, Buonassisi T, Nobre A (2018) Urban haze and photovoltaics. Energy Environ Sci 11:3043–3054
2. Nobre AM, Karthik S, Liu H, Yang D, Martins FR, Pereira EB, Peters IM (2016) On the impact of haze on the yield of photovoltaic systems in Singapore. Renew Energy 89:389–400
3. Wang H, Meng X, Chen J (2019) Effect of air quality and dust deposition on power generation performance of photovoltaic module on building roof. Build Serv Eng Res Technol 41(1):73–85
4. Council of European Union. Directive (EU) 2018/2001 of the European Parliament and of the Council on the promotion of the use of energy from renewable sources, 2018
5. Bódis K, Kougias I, Jäger-Waldau A, Taylor N, Szabó S (2019) A high-resolution geospatial assessment of the rooftop solar photovoltaic potential in the European Union. Renew Sustain Energy Rev 114:109309
6. Fattoruso G, Nocerino M, Sorrentino G, Manna V, Fabbricino M, Di Francia G (2020) Estimating air pollution related solar insolation reduction in the assessment of the commercial and industrial rooftop solar PV potential. In: Proceedings of the 20th international conference on computational science and its applications. ICCSA 2020—Lecture Notes in Computer Science, Springer

# In Situ Stability Test of a Small Amorphous Silicon Energy Harvesting Array Under Space Conditions

**H. C. Neitzert, G. Landi, F. Lang, J. Bundesmann, and A. Denker**

**Abstract** Amorphous silicon-based thin-film minimodules have been irradiated with 68 MeV protons up to a dose of $1 \times 10^{12}$ protons/cm$^2$. During the irradiation, the solar cell current under short circuit conditions, due to the photogeneration of charge carriers by the low-intensity room light and the radiation-induced generation of charge carriers, has been measured. Whereas the degradation of the photo-induced current can be continuously monitored during the experiment, the smaller radiation-induced current is only visible in current discontinuities at the beginning and the end of the radiation period. In our experiment, we measured a very similar decrease in the photo- and the radiation-induced current, both due to the proton irradiation. Therefore, we can infer that the degradation of the solar cells' photoelectrical properties is mainly due to the degradation of the amorphous silicon active material and only to a smaller content to the glass substrate's optical transmission properties. Directly after irradiation, we observed a continuous recovery of the photo-induced current, due to the room-temperature annealing of the electronic defects created in the amorphous silicon absorber layer.

**Keywords** Amorphous silicon · Proton irradiation · Radiation induced current

H. C. Neitzert (✉) · G. Landi
DIIn, Salerno University, Via Giovanni Paolo II 132, 84084 Fisciano, Italy
e-mail: neitzert@unisa.it

*Present Address:*
G. Landi
ENEA, CR-Portici, P.le E. Fermi 1, 80055 Portici, Italy

F. Lang
Cavendish Laboratory, Cambridge University, JJ Thomson Avenue, Cambridge CB3 0HE, UK

J. Bundesmann · A. Denker
Protons for Therapy, Helmholtz Zentrum Berlin, Hahn-Meitner-Platz 1, 14109 Berlin, Germany

A. Denker
Department for Mathematics, Physics and Chemistry, Beuth Hochschule, Luxemburger Straße 10, 13353 Berlin, Germany

© The Author(s), under exclusive license to Springer Nature Switzerland AG 2021
G. Di Francia and C. Di Natale (eds.), *Sensors and Microsystems*,
Lecture Notes in Electrical Engineering 753,
https://doi.org/10.1007/978-3-030-69551-4_19

# 1 Introduction

Besides applications for terrestrial large-scale energy production, amorphous silicon-based solar cells have also been proposed for space solar cells [1] or as an active material for radiation detectors [2, 3]. Generally, it has been reported that amorphous silicon has a much better radiation tolerance to energetic ions than crystalline silicon [4]. It has also been found that in a micromorph thin-film tandem solar cell, under proton irradiation the amorphous top cell is more stable, than the microcrystalline bottom cell [5]. An interesting application for small area solar cells in space could be the power-supply of autonomous sensors. Amorphous silicon has also been reported as an active material for alpha- and beta-voltaic converters [6]. Here, we test the possibility of using commercial low-cost hydrogenated amorphous silicon minimodules for low-power applications in space. For this purpose, we continuously monitored the short-circuit current of the photovoltaic cell array during irradiation and for a prolonged period after that to understand the kinetics of the degradation and an eventual recovery of the radiation-induced damage after the end of the irradiation.

# 2 Experimental and Results

## 2.1 Experimental

A series of commercial hydrogenated amorphous silicon (a-Si:H) minimodules (see Fig. 1) with an active area of 8.7 cm$^2$ has been exposed to 68 MeV protons at the HZB ion irradiation facility [7]. The a-Si:H minimodules have been irradiated with

**Fig. 1** Photo of the investigated commercial a-Si:H minimodules

a widened proton ion beam, that covered the whole active area of the array homogeneously. In this way, we were able to measure the photocurrent under small intensity room light illumination conditions and the radiation induced current. Because of the series connection of the sub-cells in the solar cell array, all single cells have to be illuminated and irradiated simultaneously to measure a substantial photocurrent and radiation-induced current.

During the in situ measurements, the minimodules have been illuminated just with room light with a constant intensity. For the characterization of the device optoelectronic properties some weeks after the irradiation, an LED-based white lamp, covering only the visible optical spectrum, has been used for illumination. Electrical measurements were done with a Keithley type "2400" source-measurement-unit.

## 2.2 Results and Discussions

The in situ monitoring results, obtained as described above, are shown in Fig. 2, where the a-SiH array short circuit current is displayed as a function of time before, during, and after proton irradiation. The proton irradiation "start" and "stop" are indicated by the arrows. Before starting the irradiation, we have a constant current value, with a low variation. This also demonstrates the low noise level obtained for these measurements in an electrically rather noisy environment (due to electrical pumps, room-light illumination, computers, and a variety of electrical measurement instruments). When the proton irradiation starts, the current monotonically decreases, until the end of the irradiation; When the proton beam stops, the current monotonically increases. The noise level of the current is slightly higher in the recovery period than the noise level before irradiation. As reported in the literature, the electrical noise

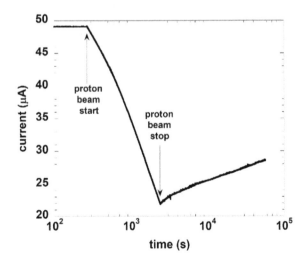

**Fig. 2** Monitoring of the current before, during and after the irradiation with 68 MeV protons of a room light illuminated a-Si:H minimodule (the beginning and the end of the irradiation period are indicated by arrows)

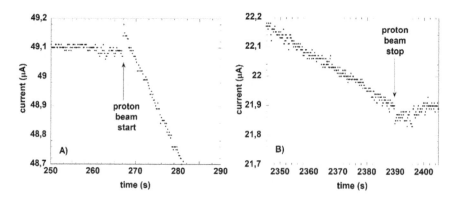

**Fig. 3** Zoom into the **a** irradiation start and **b** irradiation stop periods of the short circuit current monitoring trace during the irradiation with 68 MeV protons of a with room light illuminated a-Si:H minimodule

spectra are a good indicator of device reliability [8]. In the case of crystalline silicon solar cells, it has been shown that the temperature and bias dependent noise spectral characteristics can be used to extract the electronic defect distribution influenced by high energy particle irradiation and that the defect density is directly related to the noise amplitude [9].

Because the value of the irradiation time of 1200 s was relatively short in comparison to the value of the current monitoring period after the termination of the irradiation (8000 s), we used a logarithmic scale for the time axis. The cumulative proton dose during the irradiation was $1 \times 10^{12}$ p$^{+}$/cm$^{-2}$. We observed a monotonous decrease of the monitored current from a value of 49.1 µA before irradiation to a value of 21.9 µA after irradiation; This corresponds to around 55% decrease in the short circuit current after the proton irradiation period (Fig. 3).

Some weeks after the irradiation, the sample has been characterized again under dark and white LED-light illumination. As an example of the electric characteristics change due to the irradiation in Fig. 4, the forward dark current-characteristics of the irradiated array with a non-irradiated array are shown. In Table 1 we reported the relative change in the irradiated solar array parameters, as compared to a non-irradiated reference sample. For the latter characterization, the two arrays have been illuminated at low intensity with a white LED lamp, emitting only in the visible spectral range.

Regarding the forward dark current-voltage characteristics (Fig. 4), it can be observed that the solar cell array series resistance increased while the array diode shunt resistance decreased. These are two factors that strongly influence, for example, the solar array fill-factor and hence also the maximum value of the generated electrical power under a given illumination condition.

All measured solar array parameters (see Table 1) decreased substantially. Mainly the fill-factor (−24%) and the short-circuit current (−34%) decrease strongly, and also a minor decrease of the open-circuit voltage (−8.4%) is observed. Overall a 54%

**Fig. 4** Comparison of the forward dark current-voltage characteristics of the irradiated array (dotted line) as compared to a non-irradiated reference array

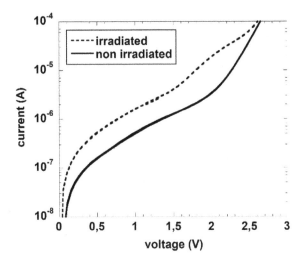

**Table 1** Photovoltaic parameter changes: open-circuit voltage (Voc), short circuit current (Isc), fill-factor (FF) and maximum electrical power (Pmax) under low-intensity LED light illumination, of the irradiated a-Si:H minimodule in comparison to a non-irradiated reference sample

|  | Voc | Isc | FF (%) | Pmax |
|---|---|---|---|---|
| Irradiated sample | 2.74 V | 156 μA | 71 | 301 μW |
| Non-irradiated reference sample | 2.51 V | 103 μA | 54 | 139 μW |
| Relative change | −8.4% | −34% | −24 | −54% |

decrease of the electrical power, obtainable under maximum power point conditions, is the result of the proton irradiation with a dose of $1 \times 10^{12}$ protons/cm$^2$. This is a stronger decrease than reported for amorphous silicon single cells, deposited, however, on "radiation resistant" glass substrates [10]. One reason could be that the enhanced short wavelength absorption due to color centers in the transparent substrate [11] is partly responsible for the short circuit current decrease. However, the relative radiation induced current decrease has roughly the same value as the photocurrent decrease, indicating that the primary degradation is related either to the active layer decrease or the degradation of the interfaces, that play for an integrated series-connected array an important role.

The measurements, to which the results in Table 1 refer, have been done some weeks after the irradiation experiment due to radiation safety regulations. There should have been, therefore, already a notable additional recovery of the array properties because of the before revealed thermal annealing process. The relative decrease of the short circuit current after some weeks is with −34% effectively still lower than the change after the in situ current monitoring (Fig. 2) with a value of −55%. It would be interesting to check how amorphous silicon solar cell arrays perform under conditions possible to find in specific space missions with a relatively high operating temperature. In this case, the balance between particle induced degradation and temperature dependent annealing could still shift more in favor of the latter mechanism. It should, however, be mentioned that amorphous silicon-based solar cells face a strong competition regarding their use for space applications from other thin-film technologies nowadays. For example, Perovskite-based single devices [12] and Perovskite/CIGS Tandem devices [13], that start already with higher initial efficiency, have been demonstrated to possess a good radiation hardness. Also, in the case of Perovskite solar cells, a partial recovery of the short-circuit-current after the end of the irradiation period has been seen during an in situ monitoring experiment [11].

# 3 Conclusions

A commercial photovoltaic minimodule, based on an array of integrated series connected amorphous hydrogenated silicon solar cells, has been exposed to high energy protons with 68 MeV energy. During irradiation, the light-induced photocurrent has been continuously monitored. After the exposure to a dose of $1 \times 10^{12}$ protons/cm$^2$, a photocurrent decrease to 55% of the initial value has been found, as well as an immediate starting recovery process. Also, a radiation-induced current decrease of 50% has been detected.

# References

1. Kuendig J, Goetz M, Shah A, Gerlach L, Fernandez E (2003) Thin film silicon solar cells for space applications: Study of proton irradiation and thermal annealing effects on the characteristics of solar cells and individual layers. Sol Energy Mat Sol Cells 79:425–438
2. Schwartz R, Braz T, Sanguino P, Ferreira P, Macarico A, Vieira M, Marques CP, Alves E (2004) Degradation of particle detectors based on a-Si:H by 1.5 MeV He4 and 1 MeV protons. J Non-cryst Solids 338–340:814–817
3. Wyrsch N, Ballif C (2006) Review of amorphous silicon based particle detectors: the quest for single particle detection. Semicond Sci Technol 31:103005
4. Kishimoto N, Amekura H, Kona K, Lee CG (1998) Radiation resistance of amorphous silicon in optoelectronic properties under proton bombardment. J Nucl Mat 258–263:1908–1913
5. Neitzert HC, Labonia L, Citro M, Delli Veneri P, Mercaldo L (2010) Degradation of micromorph silicon solar cells after exposure to 65 MeV protons. Phys Stat Sol (c) 7:1065–1068
6. Deus S (2000) Alpha- and betavoltaic cells based on amorphous silicon. In: Proceedings of the 16th European photovoltaic solar energy conference, 1–5 May Glasgow, pp 218–221
7. Röhrich J, Damerow T, Hahn W, Müller U, Reinholz U, Denker A (2012) A Tandetron™ as proton injector for the eye tumor therapy in Berlin. Rev Sci Instrum 83:02B903
8. Jones BK (1993) Electrical noise as a measure of quality and reliability in electronic devices. Adv Electron Electron Phys 87:201–257
9. Landi G, Barone C, Mauro C, Neitzert HC, Pagano S (2016) A noise model for the evaluation of defect states in solar cells. Sci Rep 6:29685
10. Klaver A (2007) PhD thesis, Delft University

11. Lang F et al (2016) Radiation hardness and self-healing of perovskite solar cells. Adv Mat 28:8726–8731
12. Lang F et al (2019) Efficient minority carrier de-trapping mediating the radiation hardness of triple-cation perovskite solar cells under proton irradiation. Energy Environ Sci 12:1634–1647
13. Lang F et al (2020) Proton radiation hardness of perovskite tandem photovoltaics. Joule 4:1054–1069

# Wireless Sensor for Monitoring of Individual PV Modules

S. Daliento and P. Guerriero

**Abstract** The paper deals with a module-level monitoring and diagnostic system consisting in wireless self-powered sensors installed on individual PV modules and performing real-time measurements of operating voltage and current, open-circuit voltage, and short-circuit current. A disconnection system assures that the PV sensor does not affect the behavior of the string during the measurement phase and allows many benefits like the automatic detection of bypass events. An experimental campaign is performed to prove the reliability and usefulness of the sensor for monitoring of PV plants. The capability to detect faults and to accurately localize malfunctioning modules in a PV string is highlighted.

**Keywords** Monitoring and diagnostics · PV module · Wireless sensor

## 1 Introduction

Nowadays, in order to ensure the return of investments, prosumers need that their Photovoltaic (PV) plants constantly operate in full efficiency, minimizing yield degradation due to faults, subsequent plant stops and production losses. To this aim, the PV market focuses its attention on innovative end effective monitoring and diagnostic approaches, ensuring accurate detection of faults and cost-effective operation and maintenance (O&M).

It should be remarked that a PV system is composed by a large number of interconnected elementary blocks (i.e., the PV modules, connected in series to form strings, then, where needed, connected in parallel to create arrays), and its performance can be even impacted by the failure of only one of them. Moreover, modules can be affected by several specific faults (e.g., break of bypass diodes, presence of dirt or

Originally submitted to Workshop AISEM 2020/ENEA-Portici.

S. Daliento (✉) · P. Guerriero
Department of Electrical Engineering and Information Technology, University of Naples Federico II, Naples, Italy
e-mail: daliento@unina.it

© The Author(s), under exclusive license to Springer Nature Switzerland AG 2021  139
G. Di Francia and C. Di Natale (eds.), *Sensors and Microsystems*,
Lecture Notes in Electrical Engineering 753,
https://doi.org/10.1007/978-3-030-69551-4_20

bird drops, cracks, delamination, and browning of solar cells, hot-spots occurrence), each of them presenting a specific effect in terms of reliability, safety, and production losses.

In general, to allow the application of modern smart maintenance practices, a monitoring and diagnostic method should be able to detect different kind of faults, accurately localize the malfunctioning modules among thousands, and to provide a precise estimation of the energy loss.

An accurate localization of defects and faults in PV plants can be achieved by using thermographic/visual inspection [1] and time domain reflectometry technique [2] during their operation and maintenance. Unfortunately, the above strategies are not suitable for real-time monitoring; moreover, they cannot quantify the energy loss due to each issue and, consequently, the returns of investments.

Better results in term of localization of faults and estimation of yield degradation can be reached by adopting real-time "high granularity" approaches relying on sensors applied to individual PV panels [3–10].

The aim of this paper is to present a monitoring and diagnostic method based on an improved version of the sensor proposed in [11]. It performs real-time measurements of operating voltage and current, open-circuit voltage, and short-circuit current of string-connected photovoltaic (PV) module (referred as *host* in the following). In particular, the proposed sensor acts as an electronic frame for the host module and offers the following features: (i) accurate detection and localization of photogenerated current degradation issues by means of continuous short-circuit-current monitoring; (ii) automatic detection and localization of bypass events; (iii) detailed yield mapping by analyzing the operating points of individual panels; (iv) reliable real-time estimation of the maximum producible power of a string; (v) quantification of the yield improvements that could be achieved with a proper maintenance/upgrading, thus allowing easy comparison between investments and revenue; (vi) solution to reliability and security issues.

The paper is organized as follows: in Sect. 2, the sensor is described; in Sect. 3, the system architecture is presented; in Sect. 4, the diagnostic approach is discussed, while in Sect. 5 experimental results prove the effectiveness of the proposed approach. Conclusions are drawn in Sect. 6.


## 2   Sensor Architecture

The block diagram in Fig. 1 depicts the architecture of the sensor, roughly divisible in the following stages: (i) disconnection circuit, (ii) power supply, (iii) control unit (MCU), (iv) measurement circuit, and (v) wireless transceiver. In the following, for sake of clarity, main characteristics of the sensor stages will be presented. However, further details about the design of the sensor can be found in [11].

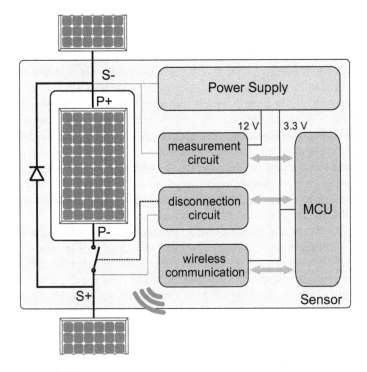

**Fig. 1** Sensor architecture

## 2.1  The Disconnection Circuit

Referring to Fig. 2, a power switch SW1 is inserted between the input terminal S+
of the sensor and the negative terminal P− of the host PV module.

**Fig. 2** Power supply stage
and disconnection circuit

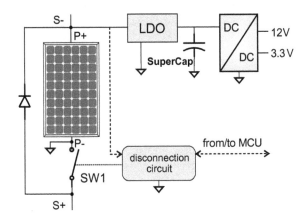

When the switch is closed, the two terminals are short-circuited and the presence of sensor is completely transparent to the PV string, not affecting the normal operation of the whole PV generator. On the contrary, when it is open, the host module is completely disconnected from the string, while the string current flows through the diode. In this case, being the sensor still in parallel to the host module, it can change operating point to perform the electrical characterization.

In general, SW1 is closed, and the MCU opens it only during measurements. However, two additional disconnection mechanisms are allowed, due to the detection of a bypass event or to an external command. In the first case, as a bypass event occurs, the disconnection circuit generates an interrupt signal. Consequently, the MCU opens the switch, thus zeroing the current in the module to avoid undesired hotspot occurrence. In the recent literature, the adoption of the disconnection stage to replace the bypass diodes has been propose [12].

In the second case, the sensor allows to intentionally bypass the host module by sending a specific command (*on-demand* disconnection). This feature was initially introduced for reliability purposes. Nevertheless, as discussed in the [13], the intentional disconnection of a module limiting the string current can prevent undesired energy yield reductions, favoring the MPPT to identify the global MPP. According to this, the proposed circuits can be considered both a sensor and an actuator.

## 2.2  The Power Supply Stage

The sensor is able to harvest energy from the host module, thus avoiding addiction power cables. The power supply stage comprises a linear voltage regulator (i.e., LDO in Fig. 2) followed by dedicated step-up voltage regulators providing the other stages with the adequate voltage levels (i.e., 3.3 V for MCU and wireless transceiver, 12 V for measurement stage). Moreover, it exploits a 2F-supercapacitor acting as Energy Storage System (ESS). In particular, the ESS is devised to feed the sensor during measurements (i.e., when the module is disconnected and the LDO is disabled). Moreover, it allows stable power supply even in case of suddenly change in the irradiation levels.

## 2.3  MCU

The sensor control unit (MCU) is realized by an 8-bit micro-controller (Microchip PIC18LF4620). As usual in wireless sensors, the MCU generally operates in sleep-mode to reduce the power consumption. As the wireless transceiver receives a measurement request, MCU awakes to handle the measurement process, send data, restore the disconnection status; finally, it falls again in sleep mode. Moreover, under bypass condition, when the switch SW1 is kept constantly open, MCU periodically awakes and checks if the bypass event is ceased to reconnect the module to the string.

## 2.4   The Wireless Communication

The communication protocol is based on Microchip MiWi [14] (IEEE 802.15.4 compliant, 2.4 GHz frequency). In the network, each node (sensor) is identified by a specific address, thus allowing the immediate localization of malfunctioning PV modules in the field. On board the sensors, the physical layer is managed a MRF24J40MA transceiver (up to 400 ft range, throughput of 250 kbps), fully compatible with the adopted MCU and allowing low power consumption (typically 19 mA in RX, 23 mA in TX, 2 μA in sleep [15]).

## 2.5   The Measurement Circuit

As described in Fig. 3, the measurement circuit mainly performs current and voltage sensing, while solid-state switches properly modify the operating point of the host module.

The measurement procedure is articulated as follows: (i) initially, SW1 and SW2 are closed, while SW3 is open (string current mainly flows through SW1); (ii) measurement of both operating current and voltage (SW1 opens to force the whole string current flowing through SW2); (2) disconnection (SW1 and SW2 are open); (3) the LDO is disabled, while the supercapacitor ensures the power supply of the sensor; (4) open-circuit voltage is acquired; (5) the module is driven into short-circuit condition by closing SW3 (SW2 is still open) and, after a dead-time, the current value is acquired; (6) the LDO is reactivated and the module is reconnected to the string (SW1 and SW2 are closed, while SW3 is open).

**Fig. 3**  Measurement circuit

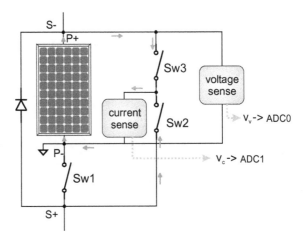

The measurement stage works properly even in bypass conditions, and the open-circuit voltage and the short-circuit current can be monitored also if the module is intentionally kept disconnected (*on-demand* disconnection).

## 3 System Configuration

In Fig. 4, a typical application is depicting. A PV string composed by N modules powers an inverter, and sensors are installed on all modules to form a wireless sensor network. At logic level, a central station manages the network, sending commands to the sensors and collecting data from the field. At physical level, specific units, referred as *coordinators,* allow the communication among the sensors and the central station, adapting 2.4 GHz wireless communication to RS485 serial bus. The proposed approach is easily scalable, thus resulting suitable for large PV plants. In fact, each coordinator handles up to 256 sensors, while the central station communicates with

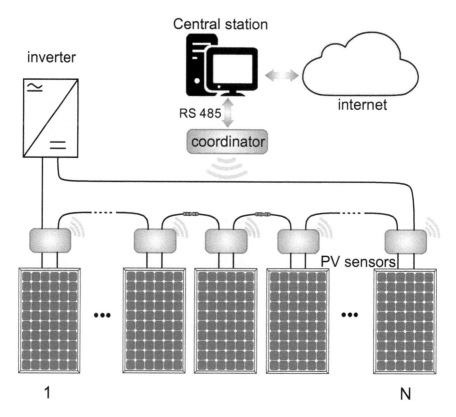

**Fig. 4** Block diagram corresponding to the *individual panel level* configuration

up to 256 coordinators on the same bus (maximum serial cable length 1000 m). Finally, the central station makes data available on the internet.

## 4 Monitoring and Diagnostic Approach

In general, a monitoring and diagnostic method should identify any fault in PV plant, accurately localize the malfunctioning module and quantify the energy losses, to justify the cost of the on-site intervention of an operator.

The proposed method exploits data collected by the sensor network to determine the performance of individual modules in their actual operating conditions with respect to the Standard Test Conditions (STC), reported in the datasheet.

For a healthy module, in fact, the maximum power $P_{MPP}$ can be derived from the value at STC, by applying a proper temperature and irradiance correction

$$P_{MPP}(G, T) = \frac{G}{G_{STC}}(1 - \gamma(T - 25\,°C))P_{MPP}(G_{STC}, 25\,°C) \qquad (1)$$

where $T$ is the junction temperature, $G$ is the irradiation level, and $\gamma$ is the temperature coefficient of the power.

Nevertheless, an indirect measure of the actual irradiance G can be performed by monitoring the short-circuit current

$$G = \frac{I_{SC}(G, 25\,°C)}{I_{SC}(G_{STC}, 25\,°C)}G_{STC} \simeq \frac{I_{SC}(G, T)}{I_{SC}(G_{STC}, 25\,°C)}G_{STC} \qquad (2)$$

while the junction temperature $T$ can be derived from the open-circuit voltage as follows

$$T = \frac{V_{OC}(G, T)}{n\frac{k}{q}\ln\left(\frac{I_{SC}(G,T)}{I_{SC}(G,T)e^{-\frac{nq}{k \cdot 25°C}V_{OC}(G_{STC},25°C)}}\right)} \qquad (3)$$

where N is the number of solar cells in the module, $n$ is the ideality factor (i.e., 1.2 in c-Si cells), q is electric charge of an electron, k is the Boltzmann constant.

As a result, proposed sensors allow a real-time comparison between the expected power and the actual, thus detecting any undesired power losses. In other words, the sensor network provides an accurate map of the losses in the PV field, thus ensuring both localization and energy quantification of faults.

Moreover, considering the most common causes of power losses in PV plants, a further analysis can be used to classify faults as follows:

1.  degradation of the module: typically due to delamination, browning, or accumulation of dust, it deals with a general degradation of the photogenerated current. It can be identified by comparing the short-circuit current corresponding to

modules with the same orientation with respect to the sun (i.e., tilt and azimuth angles).

2.  degradation of individual cells: due to bird drops or to defects in individual cells (i.e., crack, soldering issues, etc.), this kind of faults affects the capability to conduct current of the sub-modules, often leading to the activation of the bypass diode and the occurrence of hot-spot. In particular, the activation of a bypass diode causes a sudden drop of the operating voltage (i.e., 1/N of $V_{OC}$ for a module with N bypass diodes). Moreover, it should be considered that this kind of faults appears randomly and produces effect during the entire day.

3.  partial shading: due to objects close to the plant, partial shading events produce effects similar as the previous group but take place every day in the same time interval. The shadows, in fact, follow the apparent trajectory of the Sun; it results in the progressive bypass of the sub-modules, thus inducing a typical ladder shape in the operating voltage (see Sect. 5).

4.  faults of bypass diode: short-circuited bypass diode leads to a low open-circuit voltage.

5.  fault of fuses: fuses installed in the connection boxes are typically affected by open-circuit failure, easily revealed by the permanent zeroing of operating current of all the modules.

## 5  Experiments

In this section, experimental data prove that the monitoring approach based on the proposed sensor allows a deep insight of faults, thus meeting diagnostic purposes.

The experimental analysis was conducted on a PV string composed of ten 210Wp modules, each equipped with three bypass power diodes, to explore the effect of architectural shading (due to e.g., chimneys, TV antennas, poles). The experiment was performed by placing a pole in the close proximity of the string.

According to the approach proposed in Sect. 3, for each module in the string the estimated maximum power $P_{MPP}$ obtained by Eq. (1) and the measured operating power P were compared to highlight energy losses. In Fig. 5a, experimental results corresponding to module #4 are reported, this latter being representative of modules not affected by shading. A good match can be found between the two power curves, with a negligible energy loss of about 40 Wh over the whole day.

On the contrary, module #7 exhibits a significant mismatch between power curves in the time interval spreading from 13:30 and 18:00, thus resulting in about 320 Wh energy loss over the day. It suggests the occurrence of a fault.

A further analysis was performed to identify the specific source of energy loss. In Fig. 6, the open-circuit voltage, the short-circuit current, and the operating voltage and current are compared. At about 13:30, the architectural shading affects module #7. As a consequence, the three sub-modules fall progressively bypassed, while the operating voltage assumes the typical ladder shape. In fact, due to the activation of a bypass diode, in fact, the operating voltage instantaneously decreases by about

**Fig. 5** Comparison between the estimated maximum power $P_{MPP}$ and the actual operating power P (red line) for **a** module #4 and **b** module #7. The green area represents the energy losses

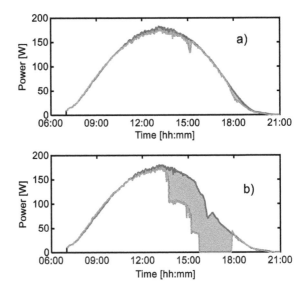

1/3 of the open-circuit voltage. In particular, at 15:00, the pole projects its shadow over the last sub-module, thus impacting the global short-circuit current (see Fig. 6a) and causing, at 15:30, the complete bypass of the module. As soon as the bypass occurs, the sensor detects the fault event and moves in disconnection mode, zeroing the module current and, then, the module power.

It is worth noting that this analysis requires that the sensors properly work under partial shading conditions. Moreover, no additional sensors (i.e., temperature and irradiation sensors) are needed to apply the proposed diagnostic strategy.

# 6 Conclusions

A wireless sensor to be employed in monitoring and diagnostics of PV plants has been presented. The sensors are installed on each individual module in the monitored PV plant to form a wireless sensor network. Moreover, being able to harvest energy from the host modules, they do not require additional cables. The proposed sensor allows the real-time monitoring of operating voltage and current, open-circuit voltage, and short-circuit current without affecting the normal operation of the plant, thanks to a specific disconnection system. As a further feature, the sensor autonomously detects bypass events and disconnects the host module to avoid hot-spots. In addition, it provides on-demand disconnection. In the paper, a specific monitoring and diagnostic method is proposed, which carries out the energy loss map of the PV plant at module-level, by comparing the power produced by each module to the maximum producible power. To estimate the latter quantity, the behavior in STC is corrected

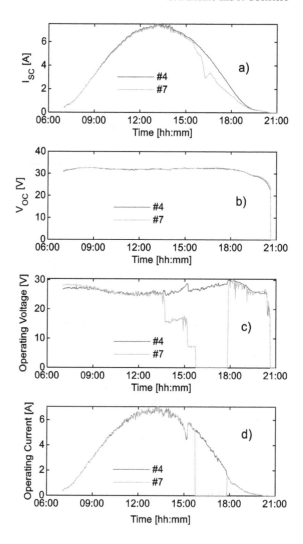

**Fig. 6** Time behavior of **a** the short-circuit current $I_{SC}$, **b** the open-circuit voltage $V_{OC}$, **c** the operating voltage, and **d** the operating current corresponding to module #4 (blue line) and module #7 (red line)

by exploiting short-circuit current and open-circuit voltage data as indirect measurement of temperature and irradiance. Experimental results evidence the capability of the proposed approach to detect faults and to accurately localize malfunctioning modules in a PV string.

# References

1. Quater PB, Grimaccia F, Leva S, Mussetta M, Aghaei M (2014) Light unmanned aerial vehicles (UAVs) for cooperative inspection of PV plants. IEEE J Photovoltaics 4(4):1107–1113

2. Takashima T, Yamaguchi J, Masayoshi I (2008) Fault detection by signal response in PV module strings. In: Proceedings of IEEE photovoltaic specialists conference, pp 1–5

3. Hanson AJ, Deline CA, MacAlpine SM, Stauth JT, Sullivan CR (2014) Partial-shading assessment of photovoltaic installations via module-level monitoring. IEEE J Photovoltaics 4(6):1618–1624

4. Resadi M, Costa S, Cesana M (2010) Control and signalling device for photovoltaic modules. European Patent Application EP2159766 A1

5. Available: http://www.spimsolar.it/

6. Ciani L, Cristaldi L, Faifer M, Lazzaroni M, Rossi M (2013) Design and implementation of a on-board device for photovoltaic panels monitoring. In: Proceedings of I2MTC, pp 1599–1604

7. Sanchez-Pacheco FJ, Sotorrio-Ruiz PJ, Heredia-Larrubia JR, Perez-Hidalgo F, Sidrach de Cardona M (2014) PLC-Based PV plants smart monitoring system: field measurements and uncertainty estimation. IEEE Trans Instrum Meas 63(9):2215–2222

8. Gargiulo M, Guerriero P, Daliento S, Irace A, d'Alessandro V et al (2010) A novel wireless self-powered microcontroller-based monitoring circuit for photovoltaic panels in grid-connected systems. In: Proceedings of IEEE international symposium on power electronics, electrical drives, automation and motion, pp 164–168

9. Guerriero P, d'Alessandro V, Petrazzuoli L, Vallone G, Daliento S (2013) Effective real-time performance monitoring and diagnostics of individual panels in PV plants. In: Proceedings of IEEE international conference on clean electrical power, pp 14–19

10. Guerriero P, Vallone G, Primato M, Di Napoli F, Di Nardo L, d'Alessandro V, Daliento S, A wireless sensor network for the monitoring of large PV plants. In: Proceedings of IEEE international symposium on power electronics, electrical drives, automation and motion, pp 960–965

11. Guerriero P, Di Napoli F, Vallone G, d'Alessandro V, Daliento S (2016) Monitoring and diagnostics of PV plants by a wireless self-powered sensor for individual panels. IEEE J Photovoltaics 6(1):286–294. https://doi.org/10.1109/JPHOTOV.2015.2484961

12. Guerriero P, Daliento S (2019) Toward a hot spot free PV module. IEEE J Photovoltaics 9(3):796–802. https://doi.org/10.1109/JPHOTOV.2019.2894912

13. Guerriero P, DI Napoli F, d'Alessandro V, Daliento S (2015) Accurate maximum power tracking in photovoltaic systems affected by partial shading. Int J Photoenergy, Article number 824832

14. AN1066 MiWi Wireless Networking Protocol Stack—Application note available on www.microchip.com

15. MRF24J40MA RF Transceiver Module—Technical Data Sheet available on www.microchip.com

# Quantum Computing Meets Artificial Intelligence

G. Acampora

**Abstract** The world of computing is going to shift towards new paradigms able to provide better performance in solving hard problems than classical computation. In this scenario, quantum computing is assuming a key role thanks to the recent technological enhancements achieved by several big companies in developing computational devices based on fundamental principles of quantum mechanics: superposition, entanglement and interference. These computers will be able to yield performance never seen before in several application domains, and the area of artificial intelligence may be the one most affected by this revolution. Indeed, on the one hand, the intrinsic parallelism provided by quantum computers could support the design of efficient algorithms for artificial intelligence such as, for example, the training algorithms of machine learning models, and bio-inspired optimization algorithms; on the other hand, artificial intelligence techniques could be used to reduce the effect of quantum decoherence in quantum computing, and make this type of computation more reliable. This position paper aims at introducing the readers with this new research area and pave the way towards the design of innovative computing infrastructure where both quantum computing and artificial intelligence take a key role in overcoming the performance of conventional approaches.

**Keywords** Quantum computing · Artificial intelligence · Machine learning · Optimization

## 1 Introduction

It was 1981 when Richard Feynman delivered his seminal lecture "Simulating Physics with Computers", where he observed that quantum mechanical effects, such as superposition and entanglement, could not be simulated efficiently on classical

G. Acampora (✉)
Department of Physics "Ettore Pancini", University of Naples Federico II, Naples, Italy
e-mail: giovanni.acampora@unina.it; giovanni.acampora@na.infn.it

Istituto Nazionale di Fisica Nucleare, Sezione di Napoli, Naples, Italy

G. Di Francia and C. Di Natale (eds.), *Sensors and Microsystems*,
Lecture Notes in Electrical Engineering 753,
https://doi.org/10.1007/978-3-030-69551-4_21

computers. This observation gave rise conjecture that computation could be done more efficiently if quantum effects are used to perform it [1]. Indeed, the exploitation of quantum effects in computation enables the design of quantum systems showing a high degree of parallelism, which exponentially grows with the size of the quantum system itself. This conjecture was proven in 1994 by Peter Shor, who designed an efficient quantum algorithm for factoring integers numbers [2]. Indeed, with respect to the classical computation applied to the problem of the integer factorization, there is no algorithm able to compute all factors of an integer number $n$ in a polynomial time $O\left(b^k\right)$, where $b$ is the number of bits used to represent $n$ and $k$ is a constant; vice versa Shor's algorithm runs in $O\left(b^3 \log b\right)$ time, and it uses $O\left(b^2 \log n \log \log n\right)$ quantum gates. A further prove of Feynman's conjecture is provided by the Grover's quantum algorithm that searches an unstructured unordered list with $N$ entries, for a marked entry, using only $O\left(\sqrt{N}\right)$ queries instead of the $O(N)$ queries required classically [3]. Shor and Grover's algorithms are only some of the most representative examples of quantum algorithms yielding better performance than their classical counterparts [4] and indeed, today, there is a growing numbers of research activities which are achieving promising results in this classical-quantum challenge. In particular, as reported by IBM website,[1] there are at least three main research areas where quantum computing can take advantage with respect to classical algorithms: chemistry, optimization, and artificial intelligence. Indeed, as highlighted by Richard Feynman, classical computers are not able to simulate natural processes and, for example, they are not able to efficiently model how biomolecules behave and interact among them at the atomic or sub-atomic level to design novel drugs. From the optimization point of view, the efficiency provided by quantum superposition and entanglement can result into the design of quantum algorithms able to navigate the solution space of hard optimization problems in a more efficient way than classical approaches; these results could strongly improve the performance of information systems in different industrial domains such as logistics, finance and energy management. Finally, methodologies from the area of artificial intelligence are particularly suitable to be quantumly improved to train and run machine learning models in better performing hard tasks such as classification, regression, clustering and so on.

In our opinion, among the above possible application of quantum computing, the meeting between quantum techniques and artificial intelligence is the one of greatest interest for two main reasons: (1) the great availability of data promptly available through high performance communication networks (optical fibers, 5G, etc.), and (2) the growing number of research activities accomplished in the area of machine learning in an indefinite number of application areas. Moreover, from the opposite point of view, quantum computing could benefit from the usage of classical artificial intelligence techniques able to support the design of quantum devices characterized by high tolerance towards decoherence phenomena. In particular, in the area of sensor networks, quantum computing and artificial intelligence could be very useful in both

---

[1]https://www.ibm.com/quantum-computing/learn/welcome.

analyzing classical and quantum states, opening completely new scenarios in the context of data analytics.

In this position paper it is provided both an overview of possible applications of quantum computation in artificial intelligence area, and a discussion about the possible usage of artificial intelligence techniques to address issues in designing quantum computers [5].

## 2    Quantum Computing-Aided Artificial Intelligence

Today, artificial intelligence is everywhere thanks to its capability of transforming raw data in sophisticated models able to provide added values to our daily tasks and support our life. From the personal point of view, our mobiles, cars, houses and so on, can be view as complex systems composed of a plethora of different devices that captures data by means of efficient internetworking frameworks and learns mathematical models able to predict humans' actions and plan a sequence of actions to support people in their activities. At the same way, factories and companies use data collected by financial applications and industrial machinery to learn new strategies for logistics, maintenance and other planning activities. Many examples could be cited to testify to the pervasiveness of artificial intelligence in our life. The strong use of these applications has opened new scenarios in the research field of artificial intelligence where new algorithms are designed for training more and more efficient models capable of dealing with ever-increasing amounts of heterogeneous data. However, in spite of this growing research activity, the design of artificial intelligence algorithms may be limited by the technological constraints imposed by current computing paradigms. Indeed, the so-called Moore's law, the observation that the number of transistors in a dense integrated circuit doubles about every two years, is going to expire and, as a consequence, there will no more chances to increase the the computational capabilities of electronic computers. There is a strong need of introducing novel computational paradigms based on a different physics support to try to bypass the limitations imposed by Moore's law expiration and open new research frontier in artificial intelligence. Quantum computing is the most serious candidate in taking on this role. In particular, quantum computing can support the design of innovative artificial intelligence methods where so-called *quantum agents* are able to collect information from the environment, modelling it as collection of potentially superposed and entangled quantum states, and evolve them in *virtual parallel universes* to speedup computation with respect to classic approaches. In our vision, thanks to quantum computing, it will be possible to use concepts such as superposition and entanglement to accelerate the training of machine learning models and improve the convergence times of optimization algorithms based on metaheuristics. For an example, about machine learning, it is emerging a research area where *quantum variational circuits*—quantum circuits that behave very much like the neural networks—mimic the tasks performed by classical machine learning models. The gates in quantum variational circuits behave like layers of a neural network, and they can be trained in different ways, achieving different potential benefits, such as:

1.  Obtaining faster results by minimizing run-time;
2.  Improvements of learning capacity: increase of the capability of associative memories;
3.  Improvements in learning efficiency: less training information or simpler models needed to produce the same results or more complex relations can be learned from the same data [6].

With respect to optimization algorithms, quantum computing can use superposition to store a subset of potential solutions of a given problem and move them towards a suitable suboptimal solution by means of quantum entanglement. In this scenario, the massive parallelism introduced by superposition in storing a huge numbers of solutions in few qubits, and the capabilities of entanglement in efficiently performing operations among "related" qubits, will result in revolutionary optimization approaches where both convergence time and accuracy of solutions will be strongly with respect to classic optimization strategies.

## 3   Artificial Intelligence-Aided Quantum Computing

Recently, major computer companies and quantum start-ups are working on the design of efficient and robust quantum devices where the effects of decoherence are minimized, so as to develop an ideal environment where running complex quantum algorithms. However, this result is difficult to pursue, due to the current physical limitations imposed by the current methodologies for the development of quantum computing devices. Thus, in this challenging scenario, artificial intelligence techniques can come in handy for trying to bypass the aforementioned physical limitations and support the design and development of efficient and robust quantum computing devices. As an example, in the Noisy Intermediate-Scale Quantum (NISQ) era, where it is expected to operate hardware with hundreds or thousands of quantum bits it will be crucial to improve our capability to map quantum circuits to quantum hardware in efficient way, so as to face intrinsic issues, such as decoherence and error rate, and ensure the development of performing algorithms involving a large number of qubits. In order to reach this goal, it will be critical to develop quantum compilers able to map quantum circuits to real hardware, so as to minimize the running time and maximize the likelihood of successful runs. However, as shown in [7], the above problem is particularly hard to solve and, as a consequence, artificial intelligence methods could represent an appropriate toolset to efficiently address it. As an example, one of the possible solutions to the above problem is provided by neural networks. Indeed, a machine learning model can be opportunely trained to learn the characteristics of quantum circuits and compute the most appropriate mapping between circuits and quantum hardware. The neural network is trained by using both features related the quantum circuits and features related to the quantum hardware where running circuits. As an example, in the first subset belong features such as the number of qubits in the circuit, the number of specific gates between pairs of qubits in the circuit, the number of quantum measurements performed for each qubit in the circuit, and so on; in the second subset belong quantum processors' calibration data provided by IBM, such as the error rate of specific gates.

This is just one example of how artificial intelligence will support the design of quantum systems and, it is not wrong to think that the number of similar applications can grow as the size of the circuits and devices that are intended to be built increase, giving to artificial intelligence a key role in this important context of high-performance computation.

# 4 Conclusions

The illustrated work shows how innovative technologies belonging to the areas of artificial intelligence and quantum computing can support each other to define the next generation of intelligent systems. In our vision, in the very near future, it will be common to work on the design of hybrid computing approaches, where quantum computation supports the design of new artificial intelligence techniques, and where artificial intelligence supports the implementation of efficient and robust quantum devices.

# References

1. Feynman R (1982) Simulating physics with computers. Int J Theor Phys 21(6/7):467–488
2. Shor PW (1994) Algorithms for quantum computation: discrete logarithms and factoring. In: Proceedings 35th annual symposium on foundations of computer science. IEEE Computer Society Press, pp 124–134
3. Grover LK (1996) A fast quantum mechanical algorithm for database search. In: Proceedings, 28th annual ACM symposium on the theory of computing, p 212
4. Montanaro A (2016) Quantum algorithms: an overview. NPJ Quantum Inf 2:15023
5. Acampora G (2019) Quantum machine intelligence. Quantum Mach Intell 1:1–3
6. Dunjko V, Briegel HJ (2018) Machine learning & artificial intelligence in the quantum domain: a review of recent progress. Rep Progr Phys 81(7)
7. Botea A, Kishimoto A, Marinescu R (2018) On the complexity of quantum circuit compilation. In: Proceedings of the 11th Annual Symposium on Combinatorial Search. AAAI Press

# ATTICUS: A Novel Wearable System for Physiological Parameters Monitoring

F. Picariello, I. Tudosa, E. Balestrieri, P. Daponte, S. Rapuano, and L. De Vito

**Abstract** This paper describes an innovative Internet-of-Medical-Things (IoMT) distributed measurement system for implementing personalized health services. The hardware design and realization of the wearable device that is embedded in a smart T-shirt are presented. Furthermore, an innovative compressed sensing method used for reducing the amount of transmitted and stored data, in the case of electrocardiogram acquisition is described, and the obtained results are discussed in terms of reconstruction quality.

**Keywords** Internet of Things · Wearable device · Compressed sensing · Electrocardiogram

## 1 Introduction

The Internet-of-Medical-Things (IoMT) paradigm is currently adopted for designing healthcare and health services that combine the reliability and safety of the typical medical devices with the dynamicity, generality, and scalability capabilities of Internet-of-Things (IoT) systems. As reported in [1], research and development on IoMT systems based on wearable health devices were motivated, on one hand, by the need of reducing healthcare costs keeping high quality services, on the other hand, by the need of shifting healthcare expenditure from treatment to prevention. According to the above-mentioned challenges, in literature, several kinds of wearable devices for implementing IoMT systems have been adopted [2]. One of the most promising wearable device categories is smart clothes because they provide more reliable and repeatable measurements than the others [2]. An IoMT distributed measurement system for cardio–surveillance, based on wearable devices embedded on a T-shirt is proposed in [3]. A microcontroller with a Bluetooth Low Energy (BLE) interface acquires the data provided by a five-lead electrocardiographic (ECG) sensor and stores them into a buffer. When the buffer is full, the recorded samples are

F. Picariello · I. Tudosa · E. Balestrieri · P. Daponte · S. Rapuano · L. De Vito (✉)
Department of Engineering, University of Sannio, Benevento, Italy
e-mail: devito@unisannio.it

© The Author(s), under exclusive license to Springer Nature Switzerland AG 2021
G. Di Francia and C. Di Natale (eds.), *Sensors and Microsystems*,
Lecture Notes in Electrical Engineering 753,
https://doi.org/10.1007/978-3-030-69551-4_22

sent to the smartphone, via BLE, and then processed by the smartphone application processor for obtaining QRS and T-wave information [4]. The acquired samples and the obtained information are transmitted by the smartphone to a healthcare center through an Internet connection. Other IoMT systems based on wearable devices are described in [5].

However, the reliability and usability of these systems still require improvements in terms of measurement accuracy, data reduction, and automatic detection and classification of anomalies. By using the data provided by several sensors (e.g. inertial measurement unit, bio-impedance sensor, ECG sensor), the IoMT system should perform data fusion algorithms for providing more accurate measurement results, for example by compensating automatically the artifacts due to the user movements that affect the ECG signals [6]. Especially in the case of continuous monitoring, the amount of transmitted and stored data by the IoMT system becomes huge as the monitoring period increases. Thus, an IoMT system should perform data compression with the aim of reducing this amount by keeping the information contained in the measured physiological quantities [7]. With the spread use of these systems, the pre-diagnosis and diagnosis steps, which are usually performed manually by specialized medical staff, are more complex and require a huge time for the data analysis. For this reason, an IoMT system must provide a Decision Support System (DSS), which processes the data provided by the sensors, and allows detecting and classifying anomalies automatically [8, 9] (e.g. arrhythmia classification from ECG signals) with the aim of reducing the specialized medical staff load.

In this paper, an innovative IoMT distributed measurement system based on a wearable device embedded on a T-shirt, which aims to address the above-mentioned challenges, is presented. The research activity presented in this paper is part of the project titled "Ambient-intelligent Tele-monitoring and Telemetry for Incepting & Catering over hUman Sustainability"—ATTICUS, supported by the Italian Ministry of University and Research. In particular, the proposed IoMT system overcomes the limitations of the existing systems reported in the literature, by: (i) allowing the acquisition of data from several sensors, thus enabling the implementation of data fusion algorithms at the wearable device level, (ii) compressing data using Compressed Sensing (CS) without degrading the measurement accuracy of the acquired signals, and (iii) implementing a distributed artificial intelligence system. After the introduction, the paper is organized as follows. Section 2 describes the architecture of the proposed IoMT system. The general architecture of the smart T-shirt, which implements the physical layer of the IoMT system, is presented in Sect. 3. Section 4 deals with the description of an innovative CS algorithm that allows reducing the amount of transmitted data by the smart T-shirt. Section 5 reports the results obtained from the experimental assessment of the performance of the proposed CS algorithm in terms of the Percentage of Root-mean-squared Difference (PRD).

## 2   The Proposed IoMT System

The general architecture of the proposed system is depicted in Fig. 1. It consists of:
(i) a smart wearable device (S-WEAR), (ii) an ambient intelligence device (S-BOX),
(iii) a Decision Support System (DSS), and (iv) a monitoring station.

The S-WEAR is a smart T-shirt that embeds sensors for: (i) ECG monitoring,
(ii) respiration rate measurement, (iii) galvanic skin response estimation, (iv) skin
temperature measurements, and (v) motor activity classification and monitoring.
Moreover, it embeds a microcontroller that acquires the measurements provided by
sensors and stores them in an SD memory card. The acquired measurements are sent
to the S-BOX via a Personal Area Network (PAN) (i.e. BLE) interface or to the DSS
via a Wireless Wide Area Network (WWAN) interface (e.g. NarrowBand-IoT).

The S-BOX acquires the measurements provided by the S-WEAR via BLE and
stores them into the memory for a reasonably long period (i.e. more than one month).
Furthermore, it performs the integration of the information provided by the S-WEAR
by applying data fusion algorithms and provides predictive analyses to detect anomalies on the acquired signals in quasi real-time mode [6]. Whether an anomaly is
detected, the S-BOX notifies with an alert the event to the DSS, via Internet, and
provides the real-time tracking of the signals related to the detected anomaly. The

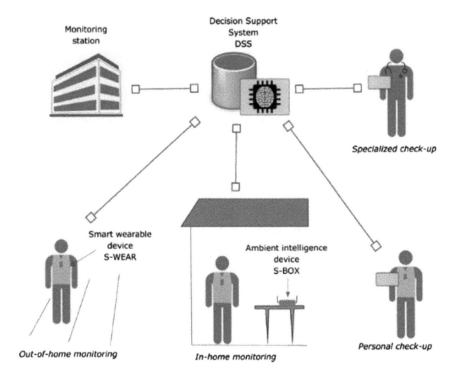

**Fig. 1**  The general architecture of the ATTICUS system [2]

data stored on the S-BOX are sent to the monitoring station through the DSS on request even in the absence of critical situations. The S-BOX is supplied by the electrical grid and is connected to the Internet via an Ethernet interface.

The DSS is a centralized system that receives the alerts notified by the S-BOX and the signals related to the detected anomalies, and according to them, it predicts emergencies automatically. If an emergency is detected, the DSS sends the related information to the monitoring station. The monitoring station is connected to the DSS via the Internet and receives from the DSS the alert messages related to each user. Furthermore, it shows the received information to the specialized user through a user interface implemented on a hand-held device or a personal computer.

## 3   The General Architecture of the S-WEAR

The S-WEAR device has been implemented according to the general architecture depicted in Fig. 2. It consists of five modules: (i) the smart T-shirt, which embeds all the electrodes used for acquiring the ECG signals, the bio-impedance (Bio-Z) measurements and the skin temperature, (ii) the core module (C), (iii) the extended ECG module (E), (iv) the position measurement module (P), and (v) the Internet interface module (I).

The core module (C) consists of: (i) the microcontroller unit (MCU) STM32WB55VGQ, which embeds the BLE transceiver integrated on-chip, (ii) the skin temperature sensor, TMP117, (Temp 1), (iii) the MAX30001, which allows acquiring one-lead ECG (ECG 1) and the Bio-Z signal, (iv) the IMU, LSM6DSL and the 3-axis magnetometer (Mag) LIS3MDL, which provide the orientation of the user, the step counting and the detection of the fall event (v) an SD memory card, and (vi) the battery and the power distribution network (PDN), which provides the correct

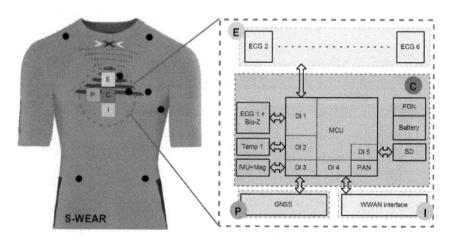

**Fig. 2** The architecture of the proposed S-WEAR device

voltage supply to all the modules. The MAX30001 (ECG 1 + Bio-Z) communicates with the MCU on the digital interface (DI 1) based on the Serial Peripheral Interface (SPI) interface. The IMU + Mag sensors communicate with the MCU through the DI 3, Inter-Integrated Circuit (I2C), and the SD memory card is driven by the DI 5 interface, SPI. The temperature sensor communicates via I2C with the MCU (DI 2). The (E) module extends the capabilities in terms of ECG monitoring of the (C) module, by providing up to five additional ECG channels. The additional ECG front ends are based on the MAX30003. The (E) module communicates with the MCU via the (DI 1) SPI interface, which is shared with the MAX30001. The MCU drives the acquisition from the different ECG channels by using the Chip Select pin associated with each sensor. Since the MCU performs the continuous monitoring of the ECG signals, a big amount of ECG data must be transmitted and then stored. Thus, the MCU performs a compression algorithm on the acquired signals before storing them on the SD card and to send them to the S-BOX.

The (P) module is adopted when it is necessary to track the position of the user in outdoor environments. It consists of a Global Navigation Satellite System (GNSS) receiver that communicates with the MCU via (DI 3) interface. The (I) module provides Internet connectivity to the S-WEAR device. This module is adopted when the S-WEAR is not able to be connected with the S-BOX via the BLE interface (i.e. out-of-home applications). In that case, the S-WEAR communicates with the DSS, directly. The (I) module is based on a WWAN interface transceiver, such as an LTE Cat NB1 transceiver that communicates with the MCU via (DI 4) interface. At this stage of the project, both (P) and (I) modules have not yet been designed and implemented.

## 4 Compressed Sensing for ECG Monitoring

For reducing the amount of data that is transmitted by the S-WEAR and stored in the S-BOX and at the DSS layer, a CS algorithm has been implemented on the MCU of the S-WEAR, as described in [10]. For one lead ECG signal compression, the implemented method performs the following steps:

1. A vector $x \in \mathbb{R}^{N \times 1}$ of $N$ discrete samples including at least one period of the ECG signal is acquired.
2. Based on this vector, the average value, $x_{avg}$, is calculated.
3. The absolute value of the point-by-point difference between the $x$ vector and its average value, $x_{avg}$, is performed, thus obtaining the vector $x_a$.
4. The vector $x_a$ is compared point-by-point with a fixed threshold value, $x_{th}$.
5. The vector $p \in \mathbb{R}^{N \times 1}$ of $N$ binary values is built as follows: (i) if the element of $x_a$ is higher than (or equal to) $x_{th}$, the value 1 is inserted in the corresponding vector index of $p$, (ii) if the element of $x_a$ is lower than $x_{th}$, the value 0 is inserted in the corresponding vector index.

6. Each row of the sensing matrix $\boldsymbol{\Phi} \in \mathbb{R}^{M \times N}$ is obtained by a circular shifting of the vector $\boldsymbol{p}^T$, where $\cdot^T$ represents the transpose operation. The number of shifted samples is equal to the compression ratio, $CR = N/M$, where $M$ is the number of compressed samples and it corresponds to the number of the $\boldsymbol{\Phi}$ rows.

7. The $M$ compressed samples, which are contained in the vector $\boldsymbol{y}$, are obtained from the multiplication between the sensing matrix $\boldsymbol{\Phi}$ and the vector $\boldsymbol{x}$.

The above operations do not require large memory capabilities and the compression algorithm exhibits a low computational load. In the ATTICUS system, the compression is performed on the MCU of the S-WEAR. The software implementation of the compression algorithm is described in [11]. To estimate the vector $\hat{\boldsymbol{x}}$ from the compressed samples $\boldsymbol{y}$, a reconstruction algorithm is required. It consists of the following steps:

1. The dyadic Mexican Hat wavelet matrix, $\boldsymbol{\Psi} \in \mathbb{R}^{N \times N+1}$, and the sensing matrix, $\boldsymbol{\Phi}$ are built.

2. The Orthogonal Matching Pursuit (OMP) algorithm is used for estimating the $R$ coefficients that represent the $\boldsymbol{x}$ vector in the domain defined by the dyadic Mexican Hat wavelet, by solving the following minimization problem: $\hat{\boldsymbol{\alpha}} = \arg\min_{\alpha} \|\boldsymbol{\alpha}\|_1$, subject to $\boldsymbol{y} = \boldsymbol{\Phi} \cdot \boldsymbol{\Psi} \cdot \boldsymbol{x}$, where $\hat{\boldsymbol{\alpha}}$ is the vector containing the $R$ estimated coefficients.

3. By multiplying the estimated $\hat{\boldsymbol{\alpha}}$ vector with the dyadic Mexican Hat wavelet matrix, $\boldsymbol{\Psi}$, the vector $\hat{\boldsymbol{x}}$ is estimated.

Being the computational load of the reconstruction higher than the compression, it should be performed on a processing platform that does not have energy consumption constraints and size limitations. For this reason, the ECG reconstruction is performed by the user device when an alert is detected, or/and when the user requires the visualization of the ECG signals. An extended version of the above-described method in the case of multi-lead ECG signals is reported in [12]. In this case, the sensing matrix, $\boldsymbol{\Phi}$, is constructed according to the ECG signal acquired by the ECG 1 module and then all the other ECG signals (i.e. ECG 2,..., ECG 6) are compressed according to it.

## 5 Experimental Results

The method presented in Sect. 4 has been implemented in MATLAB. For testing purposes, several ECG signals from the PhysioNet MIT-BIH Arrhythmia Database [13] have been used. Furthermore, the threshold value $x_{th}$ has been fixed experimentally at the 5% of the maximum of $\boldsymbol{x}_a$ for each ECG frame and the signal was framed in records of $N = 720$ samples. To evaluate the accuracy of the reconstruction, the PRD has been used as a figure of merit. The PRD is computed as follows:

**Table 1** Comparison of the PRD values obtained with the CS method proposed in [14], and the proposed CS method for the ECG signal No. 106 [10]

| Compression ratio (CR) | PRD, [14] (%) | PRD, the proposed method (%) |
|---|---|---|
| 2 | 5 | 2.4 |
| 3 | 9 | 5.5 |
| 4 | 10 | 8.6 |
| 5 | 15 | 10.9 |

$$PRD = \frac{\|x - \hat{x}\|_2}{\|x_2\|} \tag{1}$$

where, $x$ is the original signal, which contains 30 min of the acquired ECG samples, and $\hat{x}$ is the reconstructed signal. In Table 1, the comparison between the best PRD results obtained in [14], where the sparse distribution and the orthonormal Symlet-4 wavelet matrix are used, and the method proposed in this paper are reported, for the MIT-BIH Arrhythmia Database signal No: 106 [13]. The considered CR values are in the range of 2–5.

The proposed method exhibits lower PRD values for all the considered CRs. Furthermore, by considering a PRD value less than 9% that indicates a good signal reconstruction for medical applications [14], for the method reported in [14] the CR has to be chosen equal to 3 against the value of 4 of the proposed one. In addition, concerning the capabilities of tracking ECG signal variations, due to the presence of artifacts overlapped on it, another experiment has been performed using signal No: 103 of the MIT-BIH Arrhythmia Database [13]. The obtained results, which consider a CR value of 4, are presented in Fig. 3. The reconstructed signal approximates well the original signal with artifacts, in the time interval of 1090–1100 s. In [7], the method has been modified such that the sensing matrix is not evaluated in each frame, but only whether a significant change in the signal distribution is

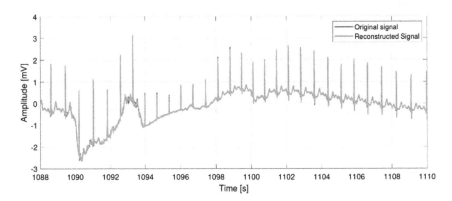

**Fig. 3** The MIT-BIH arrhythmia database signal No:103, original and reconstructed signals with artifacts $CR = 4$ [10]

found. The experimental results confirm that the method achieves better accuracy in terms of PRD than the others CS methods already available in the literature [7]. An improvement of the results in terms of PRD is obtained also in [15], where the dictionary matrix $\Psi$ is optimized according to the features in common to all the acquired records of the ECG signal of the same patient.

# 6 Conclusions

In this paper, an innovative Internet of Medical Things (IoMT) system called ATTICUS for implementing personalized health services has been presented. In particular, a description of the specific features and requirements of the wearable device that will be embedded on a smart T-shirt have been delineated and its general architecture described. An innovative CS method used for reducing the amount of transmitted and stored data in the case of ECG has been presented, and the obtained results demonstrated the goodness of the reconstruction.

Future works will be focused on: (i) the assessment of the performance of the CS algorithm by using signals acquired with the S-WEAR, (ii) the realization of the S-BOX, and (iii) the implementation of anomaly detection algorithms on the S-BOX.

# References

1. Laudato G, Rosa G, Scalabrino S, Simeone J, Picariello F, Tudosa I, De Vito L, Boldi F, Torchitti P, Ceccarelli R, Picariello F, Torricelli L, Lazich A, Oliveto R (2020) MIPHAS: military performances and health analysis system. In: Proceedings of 13th international conference on health informatics (Biostec 2020), La Valletta, Malta
2. Balestrieri E, Boldi F, Colavita AR, De Vito L, Laudato G, Oliveto R, Picariello F, Rivaldi S, Scalabrino S, Torchitti P, Tudosa I (2019) The architecture of an innovative smart T-shirt based on the Internet of Medical Things paradigm. In: Proceedings of IEEE international symposium on medical measurements and applications (MeMeA), Istanbul, Turkey
3. Trindade IJ, da Silva JM, Miguel R, Pereira M, Lucas J, Oliveira L, Valentim B, Barreto J, Silva DM (2016) Design and evaluation of novel textile wearable systems for the surveillance of vital signals. J Sens 16(1573)
4. Daponte P, De Vito L, Picariello F, Riccio M (2013) State of the art and future developments of measurement applications on smartphones. Measurement 46(9):3291–3307
5. Lymberis A, Dittmar A (2007) Advanced wearable health systems and applications. IEEE Eng Med Biol Mag 26(3):29–33
6. Hostettler R, Lumikari T, Palva L, Nieminen T, Sarkka S (2018) Motion artifact reduction in ambulatory electrocardiography using inertial measurement units and Kalman filtering. In: Proceedings of 21st international conference on information fusion (FUSION), Cambridge, UK
7. Picariello F, Iadarola G, Balestrieri E, Tudosa I, De Vito L (2021) A novel compressive sampling method for ECG wearable measurement systems. Measurement 167:1–10
8. Saadatnejad S, Oveisi M, Hashemi M (2020) LSTM-Based ECG classification for continuous monitoring on personal wearable devices. IEEE J Biomed Health Inf 24(2):515–523

9. Laudato G, Oliveto R, Scalabrino S, Colavita AR, De Vito L, Picariello F, Tudosa I (2020) Identification of R-peak occurrences in compressed ECG signals. In: Proceedings of IEEE international symposium on medical measurements and applications (MeMeA), Bari, Italy
10. Balestrieri E, De Vito L, Picariello F, Tudosa I (2019) A novel method for compressed sensing based sampling of ECG signals in medical-IoT era. In: Proceedings of IEEE international symposium on medical measurements and applications (MeMeA), Istanbul, Turkey
11. Balestrieri E, Daponte P, De Vito L, Picariello F, Rapuano S, Tudosa I (2020) A Wi-Fi IoT prototype for ECG monitoring exploiting a novel Compressed Sensing method. Acta IMEKO 9(2):38–45
12. Iadarola G, Daponte P, Picariello F, De Vito L (2020) A dynamic approach for compressed sensing of multi-lead ECG signals. In: Proceedings of IEEE international symposium on medical measurements and applications (MeMeA), Bari, Italy
13. MIT-BIH Arrhythmia Database, PhysioBank clinical database. [Online]. Available: https://www.physionet.org/content/mitdb/1.0.0/
14. Djelouat H, Zhai X, Disi MA, Amira A, Bensaali F (2018) System-on-Chip solution for patients biometric: a compressive sensing-based approach. IEEE Sens J 18(23):9629–9639
15. Picariello E, Balestrieri E, Picariello F, Rapuano S, Tudosa I (2020) A new method for dictionary matrix optimization in ECG compressed sensing. In: Proceedings of IEEE international symposium on medical measurements and applications (MeMeA), Bari, Italy

# Simulation of WSN for Air Particulate Matter Measurements

**G. D'Elia, S. De Vito, M. Ferro, V. Paciello, and P. Sommella**

**Abstract** The proposal for a wireless sensor network is recommended for continuous monitoring of particulate matter. A prototype of the Automatic air quality Measurement System (AMS), which includes a low-cost standard PM sensor, was developed as a remote node for adoption in the WSN. The results of the system calibration and the comparison with the data quality requirements of the PM measurement according to European regulations, as well as the simulation of a Smart City scenario suggests the minimum number of low-cost sensors to be used to have a system of measure deemed equivalent according to the EU regulation.

**Keywords** System calibration · WSN · IoT · Air quality monitoring systems

## 1 Introduction

Currently, one of the most serious environmental problems on the planet is given by air pollution caused by fine particles (PM10, PM2.5, with an aerodynamic diameter of approximately 10 μm and 2.5 μm respectively), also known as smog. The high concentration of these components, present in urban and industrial areas, leads to a worrying impact on diseases of the respiratory system. The European Union [1], in order to protect its citizens, has imposed the following concentration limits for PM10:

- Average annual concentration equal to 40 $\mu g/m^3$
- Average daily concentration of 50 $\mu g/m^3$

G. D'Elia · S. De Vito
CR-Portici, ENEA, P.le E. Fermi 1, 80055 Naples, Italy

G. D'Elia · M. Ferro (✉) · V. Paciello · P. Sommella
Dipartimento di Ingegneria Industriale, Università Degli Studi Di Salerno, Via Giovanni Paolo II, 132, 84084 Fisciano, SA, Italy
e-mail: mferro@unisa.it

© The Author(s), under exclusive license to Springer Nature Switzerland AG 2021    167
G. Di Francia and C. Di Natale (eds.), *Sensors and Microsystems*,
Lecture Notes in Electrical Engineering 753,
https://doi.org/10.1007/978-3-030-69551-4_23

Currently, fixed environmental monitoring stations, located in urban areas and equipped with high precision equipment, are used to verify compliance with these limits.

The creation of environmental monitoring stations involves rather high costs, mainly due to the maintenance of the high precision instruments used; this, combined with the disadvantage of not being able to perform measurements in real time, has supported the development of alternative instruments.

We propose an automatic measurement system based on distributed low-cost sensors, in particular we propose to create a low-cost wireless network,whose sensors cooperate [2] with each other to provide measurements with such uncertainty as to be accepted to verify compliance with the limits.

These spatially diffused sensing systems will allow citizens to monitor the quality of the surrounding atmosphere and, when necessary, take measures to protect their health and well-being. Collecting data from numerous low cost sensor systems [3] will facilitate the creation of highly detailed air quality maps on a local and regional scale, enabling public administrations to develop efficient plans for quality management, control and improvement of the ambient air. In this sense, improved features are required to the main components needed to develop these systems, e.g. small size and consumption sensors for integration into very small and economical measuring systems, instrumentation and control systems with high precision and repeatability, low power consumption and dimensions with communication capabilities. Last but not least, a centralized system able to process the sensor signals and data [4, 5] and assure the prescribed ambient air data quality according to international standards [1, 6].

## 2   Proposed Approach

The authors propose to use existing telecommunication networks [7] and to install low-cost devices capable of measuring environmental pollution and communicating, in order to obtain a high-resolution three-dimensional map of air pollution.

For this purpose, we analysed the Nova Fitness SDS011 sensor, a fairly recent optical sensor for measuring air quality (Particulate Matter) developed by Inovafit, a spin-off from the University of Jinan. On the basis of the calibration data achieved from 30 sensor units characterized in the laboratory [8], we simulated the behavior of suitable WSNs (including the PM sensors) in order to:

- verify the performance of the proposed system against the requirement of European guidelines;
- define the minimum number of sensors necessary to make the network metrological equivalent to traditional systems [6].

The numerical simulations of the WSN were carried out considering, for the $i^{th}$ node ($i = 1, 2, 3, ..., 15$), a corresponding measurement equal to a random variable,

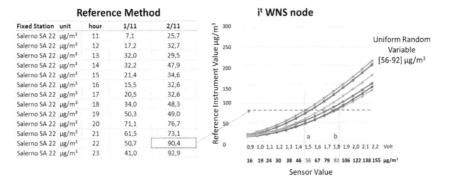

**Fig. 1** Scheme of numerical simulation: Random generation of the 24-h PM measurement by WSN node

which takes into account the values recorded from the Reference Method (RM) and the probability distribution given by sensor calibration [9].

In detail, the latter (uniform) distribution of the measurement nodes refers to the calibration results of the sensors, as shown in Fig. 1.

The guidelines described in [10, 11] are useful for assessing the feasibility of the proposed (distributed) air quality Measurement System in terms of performance and data quality.

In detail, the procedure includes the testing of two AMSs (of the same pattern, i.e. same. model, hardware, firmware and software configuration) for determining:

(i)  the between-AMS uncertainty ($u_{bs,AMS}$);
(ii) the equivalence with the Reference Method (RM) by meeting the expanded measurement uncertainty requirement.

Both the AMSs shall pass the tests if $u_{bs,AMS}$ is lower than 2.5 $\mu g/m^3$ and the expanded relative uncertainty of the results of the AMS fulfill the criterion as stated in European Directive 2008/50/EC [1] (lower than 25% in correspondence of the daily limit value $L$ equal to 50 $\mu g/m^3$). The AMSs shall be operated side-by-side with an approved implementation of the Reference Method [12] at test sites representative for typical conditions including possible episodes of high concentrations. The comparison may be performed in the form of short campaign, during which a minimum of $n = 40$ valid measurement results (each averaged over 24 h) shall be collected.

Two similar distributed AMSs (hereinafter denoted as AMS1 and AMS2) should be considered by ordering the prototypal sensor nodes according to the distance from the fixed station (operating as Reference Method) and including them alternatively between the systems. Then, a comparison, in terms of daily average for the PM10 concentration should be made, between the $n$ measurement $x_i$ obtained on the basis of the 24 h data recorded by the fixed station and the corresponding measurements $y_{i,AMS1}$ and $y_{i,AMS2}$ provided for a single 24-h period by the low cost nodes of the distributed system included respectively into the AMSs under test.

Therefore, the between-AMS uncertainty $u_{bs,AMS}$ is computed according to:

$$u_{bs,AMS}^2 = \frac{\sum_{i=1}^{n}\left(y_{i,AMS1} - y_{i,AMS2}\right)^2}{2n} \tag{1}$$

Moreover, the actual relation between the results of the distributed AMS and the (average) results of the RM is individually established for each AMS using a regression technique leading to symmetrical treatment of both variables. A commonly applied technique is the orthogonal regression that leads to the estimation of the slope $b$ and intercept $a$ of the expected linear relation:

$$y_i = a + bx_i \tag{2}$$

When the slope $b$ differs significantly from 1 and/or the intercept $a$ differs significantly from 0, the AMS shall be calibrated according to the following formula:

$$y_{i,corr} = \frac{y_i - a}{b} \tag{3}$$

Then, the orthogonal regression shall be again applied to estimate the linear relation between the (corrected) results of the distributed AMS and the average concentration by the RM:

$$y_{i,corr} = c + dx_i \tag{4}$$

Finally, the uncertainty $u_{y_{i,corr}}$ of the results by the distributed AMS is estimated according to the following formula:

$$u_{y_{i,corr}}^2 = \frac{RSS}{(n-2)} - u_{RM}^2 + [c + (d-1)L]^2 + u_a^2 + L^2 u_b^2 \tag{5}$$

where $RSS$ is the residual sum of squares resulting from the orthogonal regression, $u_{RM}$ is the random uncertainty of the Reference Method (assumed equal to $0.67\ \mu g/m^3$ when a single reference instrument is adopted), $u_a$ and $u_b$ are the standard uncertainties of the slope $b$ and intercept $a$ respectively (calculated as the square root of the corresponding variance).

In the previous formula, the first two terms represent the random uncertainty of the results of the distributed AMS whereas, the last terms are the bias at the limit value. Moreover, the covariance term between slope and intercept is not be included for simplicity.

The combined relative uncertainty $w_{AMS}$ at the relevant limit value is calculated:

$$w_{AMS}^2 = \frac{u_{yi=L}^2}{L^2} \tag{6}$$

and the expanded relative uncertainty $W_{AMS}$ is compared with the previously introduced threshold, having considered a coverage factor $k = 2$ (in view of the large number of experimental results).

## 3  Results and Discussion

We have chosen to simulate the operation of 30 sensors with comparative tests, lasting 45 days, in two distinct periods of the year.

The procedure followed to carry out the simulation was the following:

(a)  Analysing the complete data sets of the PM10 hourly ARPAC reliefs for the year 2018 on the "Salerno Via Vernieri" site during two 45-day periods. Specifically, the periods chosen are:

   1.  Winter period: from 15/11/2018 to 31/12/2018;
   2.  Summer period: from 11/04/2018 to 25/05/2018.

(b)  For each of the selected periods, the daily average values were obtained from the simulated values (according to the scheme presented above) and the daily average values of the reference system were computed from hourly PM10 datasets;

(c)  From the values obtained, the sensors were split into 2 groups of 15 units; for each subgroup an distributed AMS consisting of a variable number of sensors was considered, about which were calculated both the between-AMS uncertainty ($u_{bs,AMS}$) and the expanded relative uncertainty ($w_{AMS}$);

(d)  Proceeding with this method on each period, for each number N of sensors between 1 and 15 and repeating points b and c 30 times, 30 values of $u_{bs,AMS}$ and $w_{AMS}$ were obtained.

The graphs depicted in Fig. 2 show the trend of $u_{bs,AMS}$ and $w_{AMS}$, when an increasing number N (from 1 to 15) of sensors are included into the candidate method (i.e. each one of the distributed AMSs simulated), when the summer period is considered.

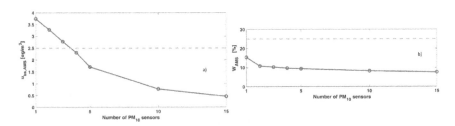

**Fig. 2** Between-AMS uncertainty $u_{bs,AMS}$ (**a**) and expanded relative uncertainty $w_{AMS}$ (**b**) of the measurement results from the distributed AMS

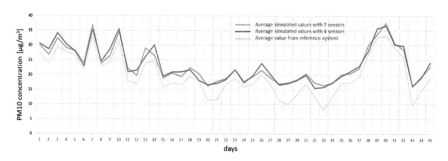

**Fig. 3** Trend of the values detected by the AMS (blue-four sensors, green-seven sensors) and the RM (orange) (color figure online)

## 4   Conclusion

The proposed distributed AMS has been simulated for two periods in which the average concentration values are different (anyway within the respect of the regulatory limit). In both cases, the distributed AMS including a number of sensors not lower than 4, results be equivalent to a measurement system according to the guidelines.

Finally, using these results, the trend of a wireless sensor network including the variable number of sensors was simulated in the 45 days following the reference summer periods (see Fig. 3 reporting measurements from the reference system and distributed AMS with N = 4 and N = 7).

A slight overestimation (about 3 $\mu$g/m$^3$) is observed from the distributed AMS including the N = 4 sensors for medium levels of the daily average PM concentration (greater 15 $\mu$g/m$^3$).

By increasing the number of sensors, the simulated trend is closer to the actual one, however, an overestimation is still observed especially for very low levels of PM10 concentration.

Further investigations will be concerned with WSN including Low-Cost sensors for other pollutants (NO$_2$, O$_3$, CO). The measurement uncertainty of new sensor types (necessary to WSN simulation) will be experimentally determined through suitable on field collocations tests.

## References

1. Directive 2008/50/EC of the European parliament and of the council of 21 May 2008 on ambient air quality and cleaner air for Europe, s.l.: s.n.
2. Fishbain B, Moreno-Centeno E (2016) Self calibrated wireless distributed environmental sensory networks. Sci Rep 6:24382. https://doi.org/10.1038/srep24382
3. De Vito S, Esposito E, Formisano F, Massera E, Auria PD, Di Francia G (2019) Adaptive machine learning for backup air quality multisensor systems continuous calibration. In: ISOEN

2019—18th international symposium on olfaction and electronic nose, proceedings, May 2019. https://doi.org/10.1109/ISOEN.2019.8823250

4. Capriglione D, Carratù M, Ferro M, Pietrosanto A, Sommella P (2019) Estimating the Outdoor PM10 concentration through wireless sensor network for smart metering. In: Baldini F, Siciliano P, Rossi M, Scalise L, Di Natale C, Ferrari V, Militello V, Miolo G, Ando B, Marletta V, Marrazza G (eds) Sensors. Lecture notes in electrical engineering. Springer, pp 399–404. ISBN: 9783030043230. https://doi.org/10.1007/978-3-030-04324-7_49

5. Colace F, Lombardi M, Pascale F, Santaniello D (2018) A multilevel graph representation for big data interpretation in real scenarios. In: 2018 3rd international conference on system reliability and safety (ICSRS). IEEE, pp 40–47

6. UNI EN 16450 (2017) Ambient air—automated measuring systems for the measurement of the concentration of particulate matter (PM10; PM2.5)

7. Ferro M, Di Leo G, Liguori C, Paciello V, Pietrosanto A (2019) Smart devices and services for Smart City. In: Proceedings of the 52nd Hawaii international conference on system sciences. 978-0-9981331-2-6. https://doi.org/10.24251/hicss.2019.156

8. Rajasegarar J, Zhang P, Zhou Y, Karunasekera S, Leckie C, Palaniswami M (2014) High resolution spatio-temporal monitoring of air pollutants using wireless sensor networks. In: Proceedings of IEEE 9th ISSNIP, Singapore

9. Carratù M, Ferro M, Paciello V, Pietrosanto A, Sommella P (2018) A smart wireless sensor network for PM10 measurement. In: IMEKO XXII World Congress 2018

10. EN 12341:2014 (2014) Ambient air-Standard gravimetric measurement method for the determination of the PM10 or PM2.5 mass concentration of suspended particulate matter. CEN

11. Guide to the demonstration of equivalence of ambient air monitoring methods (2010)

12. Standard EVS-EN16450:2017 (2017) Ambient air—automated measuring systems for the measurement of the concentration of particulate matter (PM10; PM2.5)

# Study and Characterization of LCPMS in the Laboratory

**L. Barretta, E. Massera, B. Alfano, T. Polichetti, M. L. Miglietta, P. Maddalena, F. Foncellino, F. Formisano, S. De Vito, E. Esposito, P. Delle Veneri, and G. Di Francia**

**Abstract** Atmospheric particulate matter or PM is one of the pollutants certainly most considered by professionals, both for the impact it has on the environment and for the impact, it has on human health. The official measurement methods for the detection of PM give information at long intervals and in pre-established sites giving a poor resolution from this point of view. For years, sensor manufacturers have tried to address this shortcoming with new portable devices, which have a very low sampling time (a few seconds) and a low price called the Low-Cost PM Sensor (LCPMS). This work shows a comparison of the performance of an LCPMS set. The test is carried out in the laboratory where a test chamber has been created capable of providing a controlled environment in which to test the devices.

**Keywords** Particulate matter · Air pollution · Low-cost PM sensor

## 1 Introduction

Today, the problem of environmental pollution is a delicate problem, not only for professionals but also for ordinary citizens. The government agencies responsible for environmental quality control have updated the deliveries allowed by the various environmental pollutants, in specific regulations, in works coordinated with the organizations they belong to.

Among the environmental pollutants, certainly one of the most taken into consideration, we find atmospheric particulate matter or PM. This term identifies the mass

L. Barretta (✉) · P. Maddalena
Physics Department "E. Pancini", University of Naples "Federico II", Via Cinthia 21, 80126 Naples, Italy
e-mail: Luigi.barretta3@unina.it

L. Barretta · E. Massera · B. Alfano · T. Polichetti · M. L. Miglietta · F. Formisano · S. De Vito · E. Esposito · P. D. Veneri · G. Di Francia
ENEA—Portici Research Center, P.le Enrico Fermi 1, 80055 Portici, Naples, Italy

F. Foncellino
Group STMicroelectronics, Viale Remo De Feo 1, 80022 Arzano, Naples, Italy

© The Author(s), under exclusive license to Springer Nature Switzerland AG 2021  175
G. Di Francia and C. Di Natale (eds.), *Sensors and Microsystems*,
Lecture Notes in Electrical Engineering 753,
https://doi.org/10.1007/978-3-030-69551-4_24

volumetric concentration of all the particles having dimensions of the average aerodynamic diameter of the particle itself lower than a certain threshold (measure in µm) regardless of their chemical composition, therefore the term PM10 indicates all those particles that have an average aerodynamic diameter of less than 10 µm.

Many studies [1] are aimed at assessing the effects of exposure to different particulate rates on human health. Despite the literature, this point makes us discuss, given the very general definition of PM. We can, however, distinguish on the basis of the size of the particulate, its ability to affect different areas of the respiratory tract, for example the coarse particulate with a diameter greater than 4 µm stops at the level of the trachea, the one with diameter the lower part, on the other hand, is able to penetrate the bronchi, and the alveoli when talking about ultrafine particulates (a few tenths of µm).

PM has multiple sources. We can distinguish them by categorizing according to different parameters, for example we can distinguish between natural sources (such as products of volcanic activity or marine aerosol) and anthropogenic sources (such as products of industrial activities or combustion products, etc.), between sources of coarse (corpuscles) and fine particles (fumes), or between sources indoor, which are those produced in closed environments (houses, offices, etc.), and sources outdoor, in which all the sources that emit particulates directly into the air of our cities fall.

Measuring the PM is not an easy operation, the current regulations provide for a gravimetric measurement where, by means of a filter that cuts the particles above a fixed size, the particulate is collected on a filter for 24 h and then measured on a precision microbalance with fixed temperature and humidity levels. The measurements obtained with this system are very accurate and reliable, but they bring with them two problems: (1) the measurements are obtained on daily averages and therefore it is not possible to obtain a temporally resolved information better than this sampling time; (2) Given the high cost and the number of tools, these devices are installed only in a few spatially distant sites. These two limits are in contrast with today's market demands, both as regards the working sector, one wants to think of industries or factories where for the safety of employees an indoor environmental quality control with a real-time temporal resolution or a maximum of a few seconds, or simply for home automation uses, where ordinary citizens require the control of the rate of environmental pollutants inside and outside their homes.

The answer to these problems comes from the Low Cost PM Sensors (LCPMS), which are low cost and portable optical sensors capable of performing a PM measurement. In this work we will see the evaluation of some of these sensors in the laboratory where a characterization chamber for particulates was built. In the second section, the LCPMS operating principle will be explained and a focus will be made on the characteristics of the sensors chosen to then describe the laboratory chamber and therefore the evaluation method used for the sensors. In the third section we will show and discuss the results obtained, and then move on to the conclusions in the last section of this work.

## 2   Material and Methods

LCPMS are optical sensors that exploit the scattering phenomenon of Mie to perform a count of the particles by dividing them according to their average aerodynamic diameter. Mie scattering is a very general formulation of scattering whose approximations include Rayleigh scattering when the wavelength of the incident beam (λ) is much smaller than the particle size (d), while it falls within the geometrical optics approximation when λ is much greater than d. In the operating range of the LCPMS, i.e. with λ ≥ d, the intensity of the scattered light loses its dependence on the optical parameters of the scattered particle (in particular it loses its dependence on the refractive index of the particle), which adjust a count regardless of the chemical composition of the particular being measured, when the scattering is calculated at a right angle. For this reason, the LCPMS are engineered in such a way as to have the photodetector in a direction orthogonal to the direction of the incident laser beam.

In this work we evaluate the performances of 4 LCPMS, they are the Alphasense N2 (N2), the Alphasense R1 (R1), the Novasense SDS011 (SDS011) and the Plantower PMSA003 (PMS-A003). All these sensors are capable of making a measurement for both PM2.5 and PM10, having all sampling times (settable by the user) less than 6 s. For this evaluation, a special laboratory chamber was used, built at the ENEA research center in Portici, capable of providing a controlled environment in which the concentration of the particulates can be varied, and other environmental parameters such as temperature and moisture are detected.

A particulate test chamber must be able to inject and mix the particulate in the chamber to make it homogeneous and obviously be able to measure the latter with reference instruments [2]. The chamber used (Fig. 1) is made up of a glove box where a ventilation system is used which is capable of both injecting and homogenizing the particulate matter in the chamber. The injected particulate is certified ERM-CZ100 fine dust PM10-like. The reference instrument is a Lasair III (Particle measuring systems), it is a particle counter used for monitoring clean room operations, it is able to count particles with dimensions of their average aerodynamic diameter between 0.3 and 25 μm by dividing them into 6 different channels of dimension. Once the number of particles is obtained, an indirect measurement of mass, and therefore of concentration, is obtained knowing the sampled volume, using the following [3].

$$m(d) = \frac{\pi}{6}\rho d^3 n(d) \tag{1}$$

where d is the size of the particle diameter, n is the number of particles of a given size and ρ is the mass density of the particles which is conventionally set at 1 g/cm3.

The tests were conducted with all four sensors present in the chamber simultaneously so as to have, in addition to the comparison with the reference instrument, a comparison between the LCPMS themselves. Tests were performed at three different concentration ranges: 0–40 μg/m³, 0–150 μg/m³ and 0–700 μg/m³ for the PM2.5 while for the PM10 are 0–250 μg/m³, 0-600 μg/m³ and 0–5000 μg/m³ (data relating

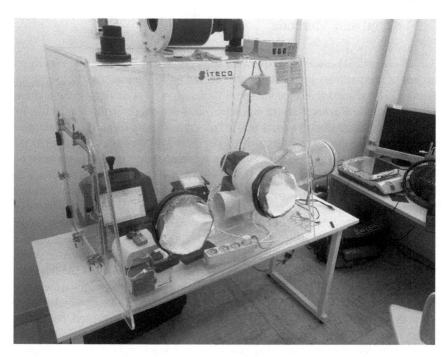

**Fig. 1** Test chamber for atmospheric particulates

to the reference instrument); we will refer to these three intervals recorded with Low Concentration (LC), Medium Concentration (MC) and Hight Concentration (HC).

## 3   Results

The comparison between the sensors and the reference instrument is done by following as parameters to test their goodness, those defined by the EPA [4]. Those taken into consideration in this work are:

- Accuracy: a measure of the overall agreement of a measurement with a known value. The Linear Regression Coefficient R2 is a measurement of an Instrument Accuracy.
- Slope: the incremental change in the response variable due to a unit change in the predictor variable.
- Bias: The systematic or persistent distortion of a measurement process that causes error in one direction.
- Measurement range: the concentration range from minimum to maximum values that the instrument is capable of measuring.

In Fig. 2 the response of the sensors (including the reference instrument) to the

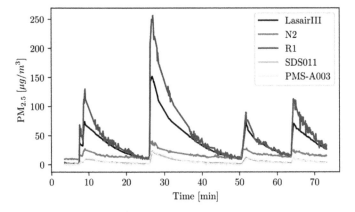

**Fig. 2** Time response of the various particulate sensors to the stimulus of PM2.5 in the medium concentration range

stimulation of the average concentrations for PM2.5 is shown. As you can see all the sensors with the exception of the R1, which suggests a different calibration for this device unlike the others including the N2 which is produced by the same manufacturer, have a significantly lower response to the stimulus, which translates into low values for slopes. In Fig. 3 a comparison between the sensors is shown not only with the reference instrument but also with each other in the medium concentration regime for PM2.5, the linear correlation coefficient is reported R2 the slope between the two sensors and the bias or better the intercept of the fit that they have one with respect to

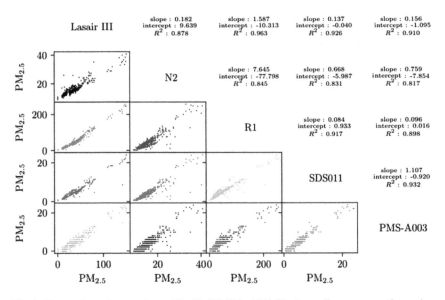

**Fig. 3** Scatter plot and comparison of the 4 LCPMS for PM2.5 in the medium concentration regime

**Table 1** Linear regression parameters between the various PMSs and the Lasair III in the different data sets (sorted by increasing concentration)

| | | N2 | | R1 | | SDS011 | | PMS-A003 | |
|---|---|---|---|---|---|---|---|---|---|
| | | PM2.5 | PM10 | PM2.5 | PM10 | PM2.5 | PM10 | PM2.5 | PM10 |
| LC | Slope | 0.178 | 0.453 | 1.366 | 4.130 | 0.114 | 0.235 | 0.058 | 0.033 |
| | Bias ($\mu g/m^3$) | 14.711 | 13.369 | −0.309 | −7.278 | 0.158 | 3.022 | −0.263 | 0.103 |
| | Accuracy | 0.551 | 0.674 | 0.908 | 0.791 | 0.872 | 0.659 | 0.607 | 0.489 |
| MC | Slope | 0.182 | 0.279 | 1.587 | 3.124 | 0.137 | 0.260 | 0.156 | 0.044 |
| | Bias ($\mu g/m^3$) | 9.639 | 8.955 | −10.313 | 6.926 | −0.040 | 9.846 | −1.095 | 3.034 |
| | Accuracy | 0.878 | 0.592 | 0.963 | 0.619 | 0.926 | 0.621 | 0.910 | 0.520 |
| HC | Slope | 0.369 | 0.321 | 2.552 | 3.008 | 0.302 | 0.485 | 0.219 | 0.079 |
| | Bias ($\mu g/m^3$) | −2.275 | 8.158 | −108.962 | 119.437 | −16.627 | 109.515 | −3.193 | 33.051 |
| | Accuracy | 0.951 | 0.901 | 0.935 | 0.905 | 0.905 | 0.793 | 0.908 | 0.744 |

another. We can summarize the results obtained in the different concentration ranges with the following in Table 1.

As you can see all the devices obtain excellent correlations for both PM2.5 and PM10 at high concentrations, the same for PM2.5 at medium concentrations. The speech at low concentrations is very different where it can certainly be said that for the PM2.5 the R1 and the SDS011 perform well but certainly not the others, worse if we consider the PM10. The reason why we see a deterioration in performance at low concentrations can be due to the fact that these devices sample very small volumes of air and therefore a large error is induced when the particles to be counted are very few as in the case of low concentrations.

The slopes are all very low, except for R1. This can be due both to a different calibration of the devices and to the nature of the particulate that is introduced into the test chamber. The question must be investigated further.

The bias, however, is always contained at low concentrations for all devices except the N2. This tells us that the N2 could give incorrect measurements below a certain threshold.

## 4 Conclusions

In this work, a comparison was made, in a special laboratory chamber capable of monitoring most of the environmental parameters as well as of injecting and homogenizing the powders in the chamber itself, on some LCPMS on which the performances were evaluated (based on the EPA parameters). It has been shown that at high concentrations the devices work very well maintaining correlation coefficients greater than

0.9. Quite the opposite occurs when the dust concentrations tend to be low where we record satisfactory results only in the case of PM2.5 for the R1 and SDS011 devices, the latter of which is very interesting given the negligible cost that is around 20$.

# References

1. Cattani G, Viviano G (2006) Stazione di rilevamento dell'istituto superiore di sanita per lo studio della qualita dell'aria: anni 2003 e 2004. Rapporti Istisan, vol 13
2. Papapostolou V, Zhang H, Feenstra BJ, Polidori A (2017) Development of an environmental chamber for evaluating the performance of low-cost air quality sensors under controlled conditions. Atmos Environ 171:82–90
3. Binnig J, Meyer J, Kasper G (2007) Calibration of an optical particle counter to provide PM2. 5 mass for well-defined particle materials. J Aerosol Sci 38(3):325–332
4. Williams R, Nash D, Hagler G, Benedict K, MacGregor I, Seay B, Lawrence M, Dye M (2018) Peer review and supporting literature review of air sensor technology performance targets

# Photometric Station for In-Vitro Diagnostic Analysis Through the Use of Organic-Based Opto-electronic Devices and Photonic Crystals

**Giuseppe Nenna, Maria Grazia Maglione, Pasquale Morvillo,**
**Tommaso Fasolino, Anna De Girolamo Del Mauro, Rosa Ricciardi,**
**Carla Minarini, Paolo Tassini, and Giorgio Allasia**

**Abstract** The present work proposes a method and an apparatus that allows the in vitro diagnostic analysis of biological samples to be carried out by analyzing the light radiation transmitted through the sample. An OLED (Organic LED) source, with broad emission spectrum for white light, was used; the optical path of the light radiation emitted by the source has been adapted to the specific wavelengths of the system (405 nm, 450 nm, 492 nm, 550 nm, 620 nm) and the detection of the signal was carried out through the use of an OPD (organic photodiode) with the appropriate photoresponse. To suitably select the analysis signal, a series of photonic crystals have been used, with the aim to also obtain a bandwidth of less than 10 nm as required by the solutions already present on the market. The proposed system gives the opportunity to reduce the overall dimensions of the analyzer compared to the systems existing on the market, avoiding the use of supplementary optical fibers and interference filters.

**Keywords** Enzyme-linked immunosorbent assay · Spectrometer · OLED · OPD · Photonic crystals

## 1 Introduction

Currently, the development of new infectious diseases requires both early diagnosis and adequate prevention and treatment. These factors increased the demand for immunological analyzers and very compact products as it has been happening in recent months with the advent of the COVID-19 emergency. In particular, serology

G. Nenna (✉) · M. G. Maglione · P. Morvillo · T. Fasolino · A. De Girolamo Del Mauro ·
R. Ricciardi · C. Minarini · P. Tassini
ENEA, CR-Portici, P.le E. Fermi 1, 80055 Naples, Italy
e-mail: giuseppe.nenna@enea.it

G. Allasia
GRUPPO FOS Srl Via Milano, 166 N/r, 16126 Genova, Italy

© The Author(s), under exclusive license to Springer Nature Switzerland AG 2021
G. Di Francia and C. Di Natale (eds.), *Sensors and Microsystems*,
Lecture Notes in Electrical Engineering 753,
https://doi.org/10.1007/978-3-030-69551-4_25

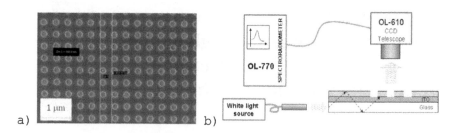

**Fig. 1** **a** SEM image of 2D PCs obtained by creating air cylinders in the ZEP polymer matrix after direct writing by EBL system respectively for PC5. **b** Example diagram of the instrumentation used for the spectral analysis of the photonic crystals

testing increased demand in order to better quantify and classify the COVID-19 cases.

This work comprises a method and an apparatus that allows the in vitro diagnostic analysis of biological samples to be carried out by analyzing the light radiation transmitted through the sample to be investigated using system parts that are not usually involved in this type of application.

In general, this type of system allows enzyme immunoassays to detect the presence of antibodies or antigens in a sample, typically in the blood, and allows, for example, to ascertain the presence of an infection. This type of instrumentation can therefore be used in tests common to this type of application such as the ELISA test (enzyme-linked immunosorbent assay) or the ELFA test (Enzyme Linked Fluorescent Assay).

Different innovative solutions have been used in the proposed system compared to the standard solutions already on the market [1]. At this stage, an OLED (organic LED) source, with broad emission spectrum for white light (Fig. 1a), was used and the optical path of the light radiation emitted by the source has been adapted to the specific wavelengths r of the system (405 nm, 450 nm, 492 nm, 550 nm, 620 nm). Moreover, the detection of the signal was carried out through the use of an OPD (organic photodiode) with the appropriate photoresponse. To suitably select the analysis signal, a series of photonic crystals (PC) have been used, with the aim to also obtain a bandwidth of less than 10 nm as required by the solutions already present on the market.

The proposed approach gives the opportunity to reduce the overall dimensions of the analyser system compared to the systems existing on the market, avoiding the use of supplementary and very long optical fibers and interference filters.

# 2 Experimental

## 2.1 OLED Fabrication and Characterization

Since the light sources must cover a very wide spectral band that includes the following wavelengths 405, 450, 492, 550 and 620 nm a white OLED was created and encapsulated by means UV-curable epoxy and a cover glass. Following the phosphorescent OLED structure is reported [2–4]:

> glass/ITO/PEDOT:PSS/NPD/SimCP/SimCP:Ir(btp)$_2$(acac)/SimCP/SimCP:
> Ir(ppy)$_3$/BCP/Alq3 /Ca/Al.

Current–voltage (I–V) and electroluminescence–voltage (EL–V) characteristics have been measured, under nitrogen and ambient temperature, using a Keithley 2400 power supply source meter with constant increment steps. Electroluminescence analysis was performed using a photodiode (Newport 818UV) connected to a Keithley 6517A Electrometer. The device electroluminescence spectra were evaluated through a CCD imaging telescope (OL610) connected to a spectroradiometer (OL770-LED).

## 2.2 OPD Fabrication and Characterization

Since the OPD of the photometric station should be sensible to the five wavelengths (405, 450, 492, 550 and 620 nm) commonly used in the immuno-enzymatic test, we selected an organic blend sensible to all the required range: a bulk-heterojunction of poly[4,8-bis(5-(2-ethylhexyl)thiophen-2-yl)benzo[1,2-$b$;4,5-$b'$]dithiophene-2,6-diyl-alt-(4-(2-ethylhexyl)-3-fluorothieno[3,4-$b$]thiophene-)-2-carboxylate-2–6-diyl)] (PTB7-Th) and [6,6]-phenyl C$_{71}$ butyric acid methyl ester ([70]PCBM). The devices with an inverted structure (glass/ITO/ZnO/PTB7-Th:[70]PCBM/MoO$_3$/Ag) were realized as elsewhere described [5] and encapsulated, like for the OLED devices, by means UV-curable epoxy and a cover glass. The device area was 1 cm$^2$ in order to avoid alignment issues.

The IV light characteristics of OPDs were performed in a nitrogen-filled glove box (O$_2$ and H$_2$O < 1 ppm) at 25 °C with a Keithley 2400 source measure unit under simulated AM 1.5G illumination provided by a class "AAA" solar simulator (Photo Emission Tech, model CT100AAA, equipped with a 150 W Xenon lamp) and its intensity was calibrated using a mono-Si reference cell with a KG5 filter for 1 sunlight intensity of 100 mW/cm$^2$. IV-dark curves were also measured by using the same apparatus.

## 2.3 PC Fabrication and Characterization

To select the wavelengths useful for the system in the immunosorbent assay area, the photonic crystals were made on an ITO coated glass substrate with 1 mm thickness and each crystal was made on a single different substrate.

A positive photoresist film (ZEP520A) with a thickness of about 200 nm was deposited by spin-coating onto the glass/ITO substrate. On the ZEP film, five photonic crystals (PC1-PC5) with the right the steps (266 nm, 296 nm, 324 nm. 362 nm and 408 nm respectively for the extracted wavelength PC1 = 405 nm, PC2 = 450 nm, PC3 = 492 nm, PC4 = 550 nm, PC5 = 620 nm) have been structured with electron beam lithography (EBL) technique (EBL-RAITH 150 system).

After being exposed to the electron beam, the positive resist ZEP520 was developed through a preliminary immersion for 90 s in xylene, then 90 s in Methyl Isobutyl Ketone (MIBK): Isopropyl Alcohol (IPA) = 1:3 and finally 30 s in IPA.

In Fig. 1a is shown the SEM image of the 2D PC obtained by creating air cylinders in the ZEP polymer matrix after direct writing by the EBL. In particular, the SEM image refers to the structures PC5 and show the structures perfectly regular and they fall within the construction limits that were initially assumed, with oscillations that do not exceed 3 nm compared to theoretical values.

The light propagating in the glass substrate which is extracted by diffraction was measured at room temperature using the setup reported in Fig. 1b which is composed by a white lamp used in Leica series microscopes, a multimode fiber, a CCD imaging telescope (OL610) and a spectroradiometer (OL770-LED) [6, 7].This apparatus has been used to analyze the optical properties of our photonic crystals in order to evaluate the resonances of the structures and the operational optical ranges.

## 3   Results and Discussion

The white OLEDs produced have been characterized electrically and optically in order to evaluate their performance. The white OLED emission spectra cover almost all the required wavelengths with three main spectral peaks located respectively at 420, 510 and 625 nm. In Fig. 2a it is possible to observe the luminance and the current density behavior of a typical white OLED device as the voltage applied varies.

In Fig. 2b, the IV-dark and IV-light (under simulated AM1.5G spectrum) curves were reported. The OPD shows a dark current of $8.56E-9$ $A/cm^2$ at $-50$ mV and under simulated AM1.5G light, the device shows a photogenerated current of $16.0E-3$ $A/cm^2$ at short circuit condition. By varying the light intensity over a range of 2 OD, the current shows a good linearity. The photodiode presents a good spectral response in the range of interest.

It seems that due to the sequence of the structures with respect to the passage of the guided light, it is recognized how the radiation emitted by the crystals includes not only that one of the crystals under evaluation but also the ones of the neighboring.

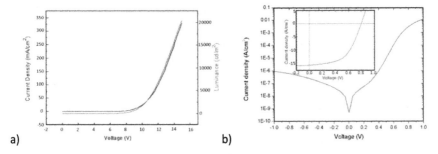

**Fig. 2 a** Current density-voltage and luminance-voltage of the white OLED. **b** Dark IV curve of a photodiode made for the photometric station, in inset the IV Light (AM1.5G)

Figure 3a shown the spectrum of the PC5 crystal, with the setup already illustrated in Fig. 1b. In this case it is recognized how the extracted spectrum is exactly the expected one. The same spectrum detected with a neutral filter with 3 Optical Density (OD) is also shown in the inset of Fig. 3a, as required by the design specifications for the systems already on the market.

The OLEDs device are made on one side of a substrate while on the other side a metallization is carried out to better confine the radiation which will be guided towards the substrate on which the photonic crystal under examination is placed. As we already stated, the OPD was used as a photodetector and, in between the crystals and the OPD, several optical filters with different OD were placed, obtaining the data reported in the table in Fig. 3b, in order to verify the linearity of this system. In general, commercial systems require linearity up to OD2 but we have tried to stress the system even further. In fact, up to OD2 there are no problems while with the last filter some problems appear related to the approach of the signal towards the background signal. Our analyzes therefore show that it is possible to select the light radiation with photonic crystals, that it is possible to have a legible signal and a linear trend up to OD2 and therefore the system proposed here can surpass

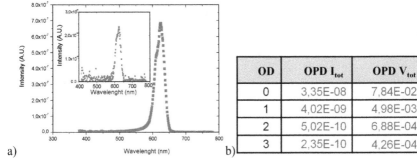

| OD | OPD $I_{tot}$ | OPD $V_{tot}$ |
|----|------------|------------|
| 0 | 3,35E-08 | 7,84E-02 |
| 1 | 4,02E-09 | 4,98E-03 |
| 2 | 5,02E-10 | 6,88E-04 |
| 3 | 2,35E-10 | 4,26E-04 |

**Fig. 3** Spectrum extracted from PC5 crystal. In the inset is reported the same spectrum detected with a 3OD neutral filter

the systems on the market thanks to improved efficiency and sizes. The problems relating to the extraction of light radiation by the crystals if illuminated in succession were highlighted and a possible alternative was studied and implemented in order to overcome this obstacle. It has been seen that one single type of crystal must be used at a time and if you want to analyze more than one sample in the 96-wells standard plate dimensions structure you could think again of optical guides with more crystals but on each guide there must necessarily be photonic crystals with the same reticular step. Considering the overall dimensions related to the used substrates to realize the devices and the photonic crystals, a total base size of about 5 cm × 10 cm is probably achieved for a complete structure for the analysis of a single sample. The height of the structure will depend, at this point, only on the size of the cuvette used and on a system for piloting the devices and reading the values, which however should not exceed 10 cm. The quoted values, concerning the dimensions of the final system, are much lower than those known on the market [8].

## 4 Conclusion

Here we present the first results related to the realization of a photometric station for in-vitro analysis of biological samples. This system is made of innovative devices like OLEDs, OPDs and photonic crystals. OLEDs have been created with a wide spectral band optimized in combination with the photonic crystals to obtain the emissive part of the station, while OPDs have been made to be used as the photodetector of a photometric station for in vitro diagnostic analysis. All the implementation steps were addressed and the problems relating to the extraction of light radiation and to the choice of assembly strategies were highlighted. Definitely, our analyzes show that it is possible to select the light radiation with photonic crystals and to have a legible signal and a linear trend up to OD2 of the samples under test., Therefore the proposed system can demonstrate better efficiency and sizes respect the ones already on the market.

**Acknowledgements** This work was carried out thanks to the projects of the Italian Public-Private partnership TRIPODE and to the ENEA Proof of Concept STADION project.

## References

1. Nenna G, Maglione MG, Morvillo P, Fasolino T, Miscioscia R, Pandolfi G, De Filippo G, Pascarella F, Minarini C, Diana R, Aprano S, Allasia G (2019) IT no. 102019000013251, 29-07-2019
2. Baldo MA, Thompson ME, Forrest SR (2000) Nature 403:750
3. Ikai M, Tokito S, Sakamoto Y, Suzuki T, Taga Y (2001) Appl Phys Lett 79:156
4. Adachi C, Baldo MA, Thompson ME, Forrest SR (2001) J Appl Phys 90:5048

5. Morvillo P, Diana R, Ricciardi R, Bobeico E, Minarini C (2015) J Sol-Gel Sci Technol 73:550–556
6. Rippa M, Capasso R, Petti L, Nenna G, De Girolamo Del Mauro A, Maglione MG, Minarini C (2015) J Mater Chem C 3:147–152 (2015)
7. Petti L, Rippa M, Capasso R, Nenna G, De Girolamo Del Mauro A, Maglione MG, Minarini C (2013) Nanotechnology 24(31)
8. https://www.accuris-usa.com/Products/microplate-absorbance-reader/

# Lab-on-Fiber Optrodes Integrated with Smart Cavities

F. Gambino, M. Giaquinto, A. Aliberti, A. Micco, M. Ruvo, A. Cutolo,
A. Ricciardi, and A. Cusano

**Abstract** A multi-responsive microgel film sandwiched between two gold layers is directly integrated on the tip of an optical fiber to form a multifunctional device able to work as an effective sensor for detecting low-molecular weight biomolecules, as well as a fiber-coupled nano-opto-mechanical-actuator triggered by light through thermo-plasmonics effects.

**Keywords** Lab-on-Fiber Technology · Plasmonics · Thermo-plasmonics · Microgels · Biosensing · Nano-Opto-Mechanical-Actuation

## 1 Introduction

The development of the Lab-on-fiber (LOF) technology has conferred to optical fibers new functionalities, thanks to the improvement of nanofabrication techniques that are allowing the effective exploitation of the nanoscale optical physics [1]. The possibility of integrating resonant nanostructures onto the tip of a standard optical fiber has led to the realization of advanced micro-sized devices, which are finding application especially in bio-chemical sensing field [2]. When properly functionalized, the nanostructures can immobilize target molecules, allowing their interaction with light localizations, thus providing signals (typically wavelength shifts of spectral features)

F. Gambino · M. Giaquinto · A. Aliberti · A. Micco · A. Ricciardi · A. Cusano (✉)
Optoelectronics Group, Department of Engineering, University of Sannio, 82100 Benevento, Italy
e-mail: a.cusano@unisannio.it

A. Ricciardi
e-mail: aricciardi@unisannio.it

M. Ruvo
Institute of Biostructure and Bioimaging, National Research Council, 80134 Napoli, Italy

A. Cutolo
Department of Electrical Engineering and Information Technology, University of Napoli Federico II, 80125 Napoli, Italy

© The Author(s), under exclusive license to Springer Nature Switzerland AG 2021     191
G. Di Francia and C. Di Natale (eds.), *Sensors and Microsystems*,
Lecture Notes in Electrical Engineering 753,
https://doi.org/10.1007/978-3-030-69551-4_26

related to the molecules concentration. In this framework, we have recently demonstrated that multi-responsive 'smart' materials can offer the potentiality for further boosting the performances of LOF biosensors, thanks to their unique capability of changing their characteristics in response to external stimuli [3]. More specifically, we have successfully integrated onto a nanostructured optical fiber tip functionalized microgels (MGs), i.e. colloidal hydrogels particles with radius ranging from few tens of nanometers up to micrometers, able to undergo reversible conformational size changes induced by external physical or chemical stimuli [4, 5]. Such variations can interact with the electromagnetic field localizations strongly tuning the resonant modes, thus giving rise to a huge amplification of the optical signal. Moreover, the MGs integration also offers the possibility of tailoring the probe response in terms of limit of detection and response time [3]. With the aim of exploiting all the degrees of freedom offered by MGs, we propose a new device consisting of a MGs film sandwiched between two gold layers in such a way to form a cavity whose length is modulated by the MGs swelling dynamics. In one configuration of this device, the bottom gold layer is patterned in such a way to excite plasmonic modes. We demonstrate that the combination of the optically resonant effects and MGs properties confers to the Cavity-enhanced LOF device the unique ability to work as a sensor for detecting small molecules with an enhanced sensitivity, as well as a nano-opto-mechanical-actuator triggered by light, setting the stage for the development of multifunctional, reconfigurable optical fiber optrodes.

## 2   Fabrication and Characterization

The device essentially consists of a MGs film sandwiched between two gold layers and integrated onto the tip of a standard single mode optical fiber (the schematic is shown in Fig. 1a). The gold layer deposited onto the fiber tip has a thickness of 30 nm and is patterned with a square lattice of holes of period 700 nm and radius

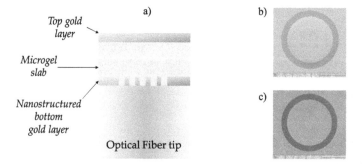

**Fig. 1** Schematic of the Cavity-Enhanced LOF device (**a**). SEM image (top view) of the probes realized with 'small' (**b**) and 'large' (**c**) MGs respectively

210 nm; this allows to excite plasmonic modes in the wavelength range of single mode optical fibers [6].

In our previous works we demonstrated that such a cavity can operate with two opposite spectral behaviors, occurring for cavity thicknesses smaller and larger than ~200 nm, which corresponds to the evanescent tail of the plasmonic resonance [7, 8]. These two regimens can be set during the fabrication step by properly choosing the MGs size, and through a careful control of the MGs deposition procedure, based on dip coating technique, optimized in our previous works [9, 10]. Therefore, we realized two different probes starting from MGs of different size, i.e. 'small' MGs (radius between ~100 nm and ~240 nm) and 'large' MGs (radius between ~190 nm and ~400 nm). Specifically, the 'small' MGs set was deposited onto the fiber tip covered by a 30 nm thick, patterned gold layer. The plasmonic nanostructure was fabricated via a focused ion beam milling process. The 'large' MGs set was instead deposited on an unpatterned gold layer of 12 nm. The two devices were then processed by depositing a 12 nm thick gold layer to form a cavity. Finally, the top mirror was patterned with a circular 5 μm width trench (external radius of 50 μm) centered in correspondence of the fiber core for ensuring a rapid and uniform MGs slab wetting. The top views of the fabricated samples are shown in Fig. 1b ('small' MGs) and 1c ('large MGs').

The two probes were placed in a temperature-controlled cuvette heated by a Peltier cells system, and characterized by exploiting the MGs thermo-responsivity for inducing the swelling dynamics, according with Dynamic Light Scattering (DLS) measurements of the MGs radius shown in Fig. 2a. The acquired spectral evolution as a function of the solution temperature is shown in Fig. 2b, c. The pseudo-color plots clearly show two opposite trends. Specifically, the cavity probe realized with 'small' MGs shows a red-shift of a reflection spectrum dip of 75.6 nm (dashed blue curve of Fig. 2b) induced by a temperature increase of 42 °C. This behavior is given by the excitation of a hybridized plasmonic mode between the two gold layers [11] that shifts towards higher wavelength in response to MGs slab shrinking induced by temperature (according with the MGs size variations shown in Fig. 2a). The probe realized with 'large' MGs shows different reflection spectral dips that blue-shifts in response to the same temperature increase. In this second case the spectral dips are given by interferometric Fabry–Perot effects occurring between the two gold layers

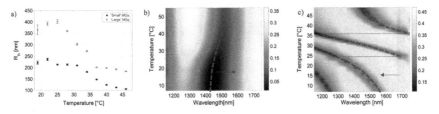

**Fig. 2** MGs radius as a function of temperature (**a**). Reflection spectra evolution as a function of solution temperature pertaining to the probe realized with 'small' (**b**) and 'large' (**c**) MGs. The insets show the measured spectra at three different temperatures

[5]. By cumulating the shifts of different spectral dips, as shown by dashed blue curve in Fig. 2c, we measured an impressive total shift of 1067 nm.

## 3   Light-Controlled LOF Actuators

The MGs thermo-responsivity combined with the plasmonic nature of the resonant modes can add new functionalities to optical fiber probes realized with the LOF technology. The gold nanostructure integrated onto the fiber tip, in fact, works as an effective local heater, whose temperature can be controlled by the input optical power, by exploiting the thermo-plasmonic effect, i.e. the local overheating caused by the light absorption [12, 13]. The light absorption by the gold film is in fact strongly enhanced by the field localization caused by the excitation of plasmonic modes. The pseudo-color plot of Fig. 3a shows the numerically evaluated absorption spectra for different MGs slab thicknesses.

In Fig. 3b we report the integral of each spectrum normalized to the integral value obtained with the standard 'open' device. It is evident that a strong absorption enhancing occurs in correspondence of a thickness of ~150 nm, i.e. when the hybridized mode is excited. By exploiting the thermo-plasmonic effect, it is thus possible to induce the MGs slab swelling/shrinking dynamics. A modulation of the input optical power, in fact, allows to control of the local overheating, and thus for the actuation of the top gold layer position. In order to demonstrate this principle, we acquired the reflection spectra of the patterned probe (realized with 'small' MGs), in correspondence of different values of the optical input power (in the range 2 μW–1 mW). During this experiment, the solution temperature was kept constant by using the Peltier cells based system mentioned above. Specifically, the same test was repeated for a solution temperature equal to 6 °C, 21° and 45°, at pH4. The plot of Fig. 4a shows the reflection spectral dip wavelength as a function of the input power level, for the three different solution temperatures. These results essentially demonstrate

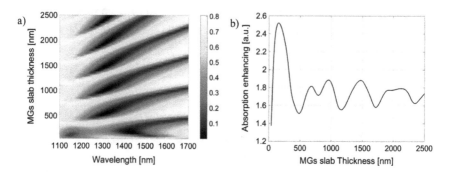

**Fig. 3** (a) Numerically evaluated absorption spectra evolution as a function of MGs layer thickness. (b) Integral absorption enhancing evaluated with respect to the standard 'open' device

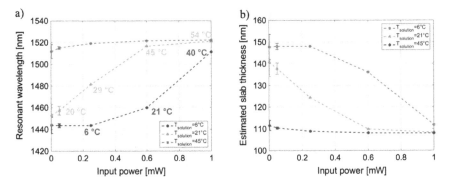

**Fig. 4** Wavelength shift (**a**) and estimated slab thickness (**b**) as a function of the input optical power

that by increasing the input optical power it is possible to induce the same wavelength red-shift observed by changing the bulk temperature solution (Fig. 2b).

The entity of the resonant shift clearly depends on buffer solution temperature, and is maximum at 6 °C, i.e. when the MGs are completely swollen. On contrary, the wavelength shifts measured at 45 °C, is negligible since MGs are already collapsed. By correlating the results of Fig. 4a and 2b, we estimated the temperature of the fiber tip as a function of the input power (cf. values reported in Fig. 4a). It is important to remark that the results shown in Fig. 2b were obtained by using a low input optical power (5 μW) in order to avoid any overheating effect. We found that the estimated temperature achieved on the fiber tip linearly increases as a function of the input power, with a slope of about 45 °C/mW. Then, to evaluate the cavity thickness variations induced by the input optical power, a numerical fitting process was implemented by taking into account the inverse relation between the RI and the thickness of the MGs slab. More specifically, the MGs slab was modeled as a uniform layer whose RI ($n_{slab}$) is given by the Eq. (1), as a function of the MGs slab thickness ($h_{slab}$), according with the model described in our previous works [14, 15],

$$n_{slab} = (n_p - n_s)K\frac{L}{h_{slab}} + n_s \qquad (1)$$

where $n_p$ and $n_s$ are respectively the RI of the NIPAM (1.47) and solution (1.33), L is the number of MGs layer (set equal to 2 according with the number of immersions during the dip coating procedure) and K is a constant depending on the MGs particle profile in dry state, measured by means of an Atomic Force Microscopy (cf. [15] for more details). The results of our analysis are shown in Fig. 4b. Coherently with the wavelength shift, the maximum cavity thickness excursion (36 nm, from 148 to 112 nm) was achieved at low temperature (6 °C). At 21 °C the estimated thickness variation range is 32 nm, (i.e. from 140 to 108 nm), while at 45 °C the actuation was negligible.

## 4 Sensitivity Enhanced Biosensors

When MGs slab is properly functionalized to respond to molecules binding events, the induced swelling/shrinking dynamics can be exploited to detect target molecules dissolved in a buffer solution, obtaining a wavelength shift that is correlated to the molecules concentration. For evaluating the capability of the Cavity-Enhanced LOF probe to detect molecules, we tested the sample working in the Fabry–Perot like regime, in order to take advantage from the huge shift observed during the preliminary thermal characterization. In the wake of our previous work, the MGs slab was functionalized with amino phenyl boronic acid (APBA) in order to make the probe sensitive to glucose molecules, chosen as benchmark for small molecules detection. The glucose assay was carried out by placing the fiber probe in a carbonate buffer solution (700 µL, pH9) whose temperature was fixed at 20 °C, containing glucose molecules at different concentrations (between 0.16 µM and 16 mM). After the insertion we monitored the reflection spectra evolution; at the end of the measurement of each concentration, the probe was regenerated in water solution for 1 h under stirring, and tested in the new concentration.

At glucose concentrations of 160 µM, 1.6 mM, and 16 mM, respectively, the sensorgram (i.e. the wavelength shifts of the reflection spectrum dip as a function of time) reached steady-state values of 58.9 nm, 242.7 nm and 1043.1 nm. These results are shown in Fig. 5a. Similarly to the above described thermal characterization, the shift obtained with the higher concentration was estimated by cumulating the single shifts of different reflection dips. Negligible wavelength shifts were instead measured for concentrations up to 16 µM. For demonstrating the enhanced sensitivity we deposited the same MGs set on a patterned device, without realizing the top gold layer. This standard MGs-assisted LOF probe (the same studied in our previous work [3]) responded to the same glucose concentrations with a blue-shift that does not exceed ~16 nm (Fig. 5b).

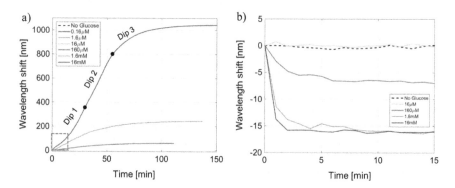

**Fig. 5** Biosensing experiment: sensorgram for different glucose concentrations for Cavity-Enhanced LOF probe (**a**) and 'standard' open device (**b**)

# 5 Conclusion

In conclusion, in this work we reported on an advanced multi-responsive LOF device arising from the integration onto an optical fiber tip of a swelling cavity composed by a MGs slab placed between two gold layers. The probe is reconfigurable for different applications, depending on the MGs characteristics and on the electro-magnetic phenomena exploited as well. By combining the thermo-plasmonic effect with the MGs thermo-responsivity we successfully demonstrated the capability of inducing changes of the top membrane position by modulating the input optical power in mW range. These results pave the way to the development of active light-triggered nano-actuators completely integrated onto the fiber tip. Moreover, by properly func-tionalizing the MGs slab, we demonstrated an enhanced sensitivity for small molecule detection up to a factor 65 with respect to a standard 'open' MGs assisted-LOF probe. Overall, the discussed results enables to achieve significant improvements in terms of performances and functionalities of LOF probes opening new avenues for the development of advanced, multifunctional, reconfigurable and tunable fiber optic devices.

# References

1. Cusano, Consales M, Crescitelli A, Ricciardi A (2015) Lab-on-fiber technology, vol 56. Springer, Heidelberg
2. Ricciardi A, Crescitelli P, Vaiano G, Quero M, Consales M, Pisco et al (2015) Lab-on-fiber technology: a new vision for chemical and biological sensing. Analyst 140:8068–8079.
3. Aliberti A, Ricciardi M, Giaquinto A, Micco E, Bobeico V, La Ferrara et al (2017) Microgel assisted Lab-on-Fiber optrode. Sci Rep 7
4. Plamper FA, Richtering W (2017) Functional microgels and microgel systems. Acc Chem Res 50:131–140
5. Wei ML, Gao YF, Li X, Serpe MJ (2017) Stimuli-responsive polymers and their applications. Polym Chem 8:127–143
6. Ebbesen TW, Lezec HJ, Ghaemi HF, Thio T, Wolff PA (1998) Extraordinary optical transmission through sub-wavelength hole arrays. Nature 391:667–669
7. Giaquinto M, Aliberti A, Micco A, Gambino F, Ruvo M, Ricciardi A et al (2019) Cavity enhanced Lab-on-Fiber technology: towards advanced biosensors and nano-opto-mechanical active devices. In: ACS Photonics
8. Ricciardi A, Aliberti M, Giaquinto A, Micco, Cusano A (2015) Microgel photonics: a breathing cavity onto optical fiber tip. In: 24th International conference on optical fibre sensors, vol 9634
9. Giaquinto M, Micco A, Aliberti A, Bobeico E, La Ferrara V, Menotti R et al (2018) Optimization strategies for responsivity control of microgel assisted Lab-On-Fiber optrodes. Sensors 18:1119
10. Scherino L, Giaquinto M, Micco A, Aliberti A, Bobeico E, La Ferrara V et al (2018) A time-efficient dip coating technique for the deposition of microgels onto the optical fiber tip. Fibers, vol 6, Dec 2018
11. Liu N, Giessen H (2010) Coupling effects in optical metamaterials. Angew Chem Int Ed 49:9838–9852
12. Tordera D, Zhao D, Volkov AV, Crispin X, Jonsson MP (2017) Thermoplasmonic semitrans-parent nanohole electrodes. Nano Lett 17:3145–3151
13. Baffou G (2017) Thermoplasmonics. World Scientific

14. Giaquinto M, Ricciardi A, Aliberti A, Micco A, Bobeico E, Ruvo M et al (2018) Light-microgel interaction in resonant nanostructures. Sci Rep 8:9331, Jun 19 2018
15. Giaquinto M, Micco A, Aliberti A, Bobeico E, Ruvo M, Ricciardi A et al (2018) Engineering of microgel assisted lab-on-fiber platforms. In: Optical fiber sensors, p TuE3

# Nanoelectrodes on Screen Printed Substrate for Sensing Cholinesterase Inhibitors

L. Iachettini and W. Vastarella

**Abstract** In this work gold and platinum nanoelectrode ensembles (NEEs) have been produced with a quite controlled pore diameter range using a polycarbonate templating membrane with pore size between 10 and 50 nm. Such nanostructured surfaces have been used in a novel hybrid electrochemical device, by coupling NEEs with disposable screen printed substrate (NEEs-SPS). Electrodes responses to hydrogen peroxide and glucose have been collected and compared with other glucose-oxidase sensors. NEEs-SPS device based on acetyl cholinesterase (AcChE) has been applied for the first time in detection of cholinesterase in form of inhibition biosensor. As a specific substrate of the main reaction acetyl thiocholine chloride salt (AcTCh$^+$Cl$^-$) has been used. Analytical performances have been evaluated in term of sensor sensitivity and linearity of response. A measurement protocol has been adopted for detecting specific organo-phosphorous compounds, using Dichlorvos as a pesticide reference. Preliminary results on fruit sample are also reported and discussed.

**Keywords** Electrochemical biosensors · Cholinesterase inhibitors · Nanoporous membranes

## 1 Introduction

Gold electrodes are largely used in electrochemical sensing in form of nanostructured materials because of their properties. Gold nanoelectrode ensembles (NEEs) can be synthesized in a huge variety of methods either by using a bottom up approach or by physico-chemical processes [1, 2]. They have been demonstrated to be a useful tool to increase the analytical performances in sensors and biosensors. By using nanoporous membranes as templates for Nano Electrode Ensembles it is possible to obtain an assembly of gold ultra-microelectrodes confined in micro-scale areas,

L. Iachettini · W. Vastarella (✉)
ENEA, CR-Casaccia, via Anguillarese 301, S. Maria di Galeria, 00123 Rome, Italy
e-mail: walter.vastarella@enea.it

© The Author(s), under exclusive license to Springer Nature Switzerland AG 2021  199
G. Di Francia and C. Di Natale (eds.), *Sensors and Microsystems*,
Lecture Notes in Electrical Engineering 753,
https://doi.org/10.1007/978-3-030-69551-4_27

which can be mainly used to immobilize specific biomolecules [3]. In many laboratories Au NEEs with a quite controlled pore diameter range have been produced and
characterized using as a template a polycarbonate membrane with pore size between
10 and 50 nm. [4]. As reported in previous works, in such a way a novel hybrid
electrochemical device was applied in biosensing, by coupling NEEs with disposable screen printed electrodes [5, 6]. Efficiency and sensitivity of the electrodes so
obtained were tested using a series a glucose oxidase ($GO_x$) nano-biosensors. Electrodes responses to hydrogen peroxide and glucose were collected and compared to
other glucose-oxidase macro-electrodes [6]. In this paper the successful deposition
of Pt nanotubes on the same template has been attained by using the same elec-
troless deposition procedure, with modifying just the last synthetic step (Pt instead
of Au plating solution) and incubation time. The comparison between NEEs-SPS
hybrid sensing system and simple screen printed sensor system is here discussed and
applied in the detection of cholinesterase inhibitors, after implementation of AcChE
biosensors. Many publication have been reported in using electrochemical biosensor
to detect the cholinesterase inhibition, due to the presence of organophosphorous
pesticides (OPs) [7, 8]. Acetyl-thiocholine chloride salt ($AcTCh^+Cl^-$) has been used
here as a specific substrate of the enzymatic reaction. Analytical performances of
such sensors and preliminary results on simplified fruit samples have been shown
and discussed.

## 2   Experimental

### 2.1   Materials and Method

Our contribution in the development of biosensors based on nanoscale material relies
on the combination between screen printed substrates and gold membranes to give a
NEE/SPS-based biosensor: as previously reported and illustrated [5, 6]. NEE/SPS-
based biosensors were tested under flow conditions, operating with a homemade
flow cell. This detection cell was provided with peristaltic pump (Gilson™ Minipuls
3) both to work under flow condition and to propel solution along a flow injection analysis system, using a 115 µL sample loop injection valve and volume
(Omnifit™, Cambridge, England). All reagents, chemicals and solvents of analytical grade were used without further purification and were from Sigma™: AcChE
from *Electrophorus electricus*, glutaraldehyde (GA) 25% solution, bovine serum
albumin (BSA, *stock solution: 4% w/v*), Dichlorvos solution, choline chloride salt
(ChCl), tiocholine salt $AcTCh^+Cl^-$. Supporting electrolyte solutions, phosphate
buffer carriers were prepared from deionized water (Synergy 185 apparatus from
Millipore™). Voltammetric and amperometric measurements, performed respectively under batch and flow conditions, were conducted with a Palm Instrument
BV™ potentiostat interface for electrochemical sensing with PalmSENS software
(2.33 version, 2008) and PS Lite 1.8. Glucose oxidase from *Aspergillus niger*, specific

activity of 198 units per mg of solid, was used as a biochemical model system to test feasibility of the sensing probe and the specific response to a target substrate after immobilization. Other instrumentation used was: Whatman 0.45 μfilter; Velp Scientifica magnetic microstirrer; Millipore™ Synergy 185 to prepare deionized water; Giro-ROCJER STR9™ Stewart basculator; Mettler analytical balance; Moll & Co. centrifuge; Hanna Instr. pH213 microprocessor pHmeter. The working flow rate (0.4 mL min$^{-1}$) and buffer pH (6.8) were fixed in all the following experiments.

## 2.2 Immobilization Techniques and Electrode Preparation

AcChE is known to be efficient in catalyzing acetyl choline chloride substrate, but also it was demonstrated that AcChE bioelectrode can generate hydrolysis of acetyl-thiocholine salt, $(CH_3)_3-N^+-CH_2-CH_2-S-CO-CH_3$ (ATCh), to give as products thiocholine (TCh) and acetic acid, after redox potential application (reaction 1). Afterwards, a rapid enzyme-less oxidation of the product TCh takes place to form the complex dithio-bis-choline according to reaction 2:

$$(CH_3)_3 - N^+ - CH_2 - CH_2 - S - CO - CH_3 + H_2O$$
$$\rightarrow (CH_3)_3 - N^+ - CH_2 - CH_2 - SH + CH_3COOH \tag{1}$$

$$2\ TCh \rightarrow dithiobischoline + 2H^+ + 2e^- \tag{2}$$

Such sequence of reactions and the resulting charge transfer on the bio-electrode is used for electrochemical sensing, after enzyme immobilization and using of ATCh as a substrate. Immobilization of AcChE was made according to different methodologies. In all the cases a preliminary procedure was adopted, by functionalization of the sensing surface with a well-known silanizing agent. Briefly, the activation is achieved by immerging the probe for 40 min in a 10% 3-aminopropyl-trietoxysilane (APTES) in 0.1 M phosphate buffer (PB) solution at pH 6.8–6.9; then the electrode is left to dry. In all the immobilization techniques, 5 μL of the resulting enzymatic solution were casted on the working electrode.

- Direct enzyme immobilization was made with freshly prepared glutaraldehyde (GA) at a concentration of 1.5% v/v in 0.1 M PB solution.
- Immobilization with crosslinking agent and covalent binding was performed with GA 1.5% v/v and 10 mg in 50 μL bovine serum albumin (BSA) of buffer mixed in a ratio of 1:4; AcChE calculated was 2104 unit μL$^{-1}$.
- Entrapment in polymeric gel of digliceryl silane (DGS), which is synthetized in our laboratories, was made in enzymatic solution 50% diluted with DGS and then dropped 5 μL on the working surface; AcChE concentration was 2500 unit μL$^{-1}$.

The inhibition of AcChE in the presence of pesticide was measured according to the stopped flow technique, i.e. by injecting different amounts of Dichlorvos diluted solutions and then stopping the flow at a fixed optimal incubation time.

## 3  Results and Discussion

### 3.1  Calibration with Different Sensor Devices

The experimental parameters were optimized choosing the best response under flow conditions. Hydrodynamic voltammogram at different voltage were reported in Fig. 1, using the biosensor with AcChE in a crosslinked matrix: ATCh concentration was fixed at 0.1 M in PB solution, keeping also constant flow rate of 0.4 mL min$^{-1}$ and buffer at pH 6.8. Unless using suitable redox mediators, in all the analyzed cases the highest current sensitivities were reached close to +700 mV of applied potential vs. Ag/AgCl pseudo-reference electrode. Hence, the applied potential of +700 mV was fixed in all the following experiments. Also pH buffer influence was tested, keeping constants enzyme and substrate concentration: highest current signals were attained between 6.9 and 7.2 (data not shown).

In Table 1 the main analytical performances are summarized for each sensor studied, differing for the immobilization methodology used. Four replicates were acquired for each substrate concentration in order to calibrate each sensor. Concentration range explored, linear dynamic range, the correlation coefficient and sensitivity (expressed as the slope of the linear calibration plot) were shown as recommended parameters. Particularly, the different sensors tested were distinguished as in the following.

Sensor 1: graphite screen printed electrode with AcChE entrapped in DGS sol–gel matrix; Sensor 2: graphite screen printed electrode with AcChE crosslinked in

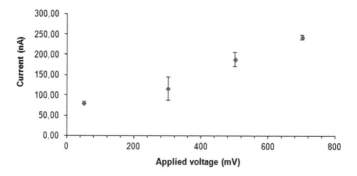

**Fig. 1** Hydrodynamic voltammogram using NEE-SPS with crosslinked AcChE, with increasing applied potential of +50, +300, +500 e + 700 mV versus Ag/AgCl pseudo-reference; substrate ATCh: 0.1 M in buffer (pH 6.9); flow rate: 0.4 mL min$^{-1}$

**Table 1** Summary of the different immobilization methods applied for each inhibition biosensor and their corresponding analytical performances

| Biosensor | Concentration range explored (mM) | Linear dynamic range (mM) | $R^2$ (linear correlation %) | Relative standard deviation | Sensitivity (nA/mM) |
|---|---|---|---|---|---|
| 1. AcChE imm. in solgel di DGS (graphite electrode) | 0.05–10 | 0.05–10 | 0.9969 | 6.7 | 152.8 |
| 2. AcChE imm. in crosslinked BSA-GA (graphite) | 0.05–10 | 0.05–1 | 0.9904 | 8.0 | 28.7 |
| 3. AcChE imm. in crosslinked BSA-GA (mediator modified graphite) | 0.05–10 | 0.1–1 | 0.9837 | 1.9 | 148.3 |
| 4. AcChE covalently bonded (graphite) | 0.02–12 | 0.02–10 | 0.9841 | 6.9 | 35.9 |
| 5. AcChE crosslinked (Au nanoelectrode on SPS) | 0.01–12 | 0.05–10 | 0.99852 | 8.5 | 1652.7 |
| 6. AcChE crosslinked (Pt nanoelectrode on SPS) | 0.01–12 | 0.05–12 | 0.9993 | 15.1 | 1280.9 |

BSA-GA matrix; Sensor 3: graphite screen printed electrode modified with Prussian Blue with AcChE crosslinked in BSA-GA matrix; Sensor 4: graphite screen printed electrode with AcChE covalently bonded; Sensor 5: gold NEE-SPS with AcChE crosslinked in BSA-GA matrix; Sensor 6: platinum NEE-SPS with AcChE crosslinked in BSA-GA matrix.

As shown in Table 1, a linear behavior was detected from 0.05 to 10 mM of ATCh concentration in PB 0.1 M, with the exception of sensor 2 and 3 where the linearity was reached up to 1 mM. Interestingly, the sensitivity resulted around tenfold higher when using nano-biosensors (biosensors 5 and 6 in Table 1) instead of simple screen printed substrate. This result proves that nanostructured electrodic materials increase the signal to noise ratio with respect to the conventional electrodes, mainly due to their higher surface active area.

Operational stability was also calculated on all the biosensors: values ranged from 7 days in case of simple entrapment until 10 days in case of enzyme covalently bonded (data not shown).

## 3.2   Cholinesterase Inhibition Measurements

In order to measure an inhibition effect, incubation phase is always required, where the sensor should be put in contact with known amounts of pesticide for a suitable time before the effective measurements. Inhibition coefficient was calculated according to relationship 3, where $I_p$ is the signal in presence of pesticide, $I_0$ is the signal without any pesticide.

$$\% \text{ inhibition} = (I_0 - I_p)/I_0 * 100 \tag{3}$$

As resulted in Fig. 2, the highest sensitivity was reached when incubation of our sample into the pesticides is 12 min, using 10 ppb Paraoxon as pesticide reference, keeping constants substrate concentration, flow rate, and applied potential (+700 mV) on sensor 2 in Table 1. Very similar curves were obtained for nanosensor 6, using 10 ppb of Dichlorvos instead of Paraoxon. When further increasing incubation more than 20 min, the process efficiency showed constant values around 29%, most probably because of a saturation effect. The standard addition procedure was applied in

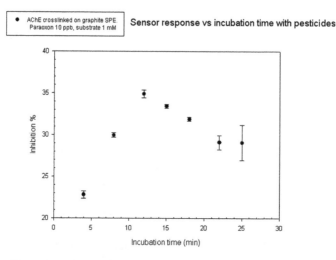

**Fig. 2** Experimental results at different incubation time of pesticide before measurements with sensor 2 of Table 1. Paraoxon concentration: 10 μg L$^{-1}$; TCh concentration: 1 mmol L$^{-1}$; + 700 mV versus Ag/AgCl pseudo-reference

a stopped flow mode, to enhance the effective inhibition of pesticide. The sensor was left for 12 min incubated with both the substrate and the fixed added amount of Dichlorvos, afterwards the stream was activated up to a current steady state.

Figure 3 shows the current profile of Pt NEE-SPS nanobioelectrode at different standard injection of Dichlorvos, according to the following steps: (A) only ATCh buffered flow, reaching the 100% of current signal; (B) sensor response after addiction of 0.5 ppb pesticide; (C) response at 1 ppb of pesticide; (D) 1.8 ppb of pesticide; (E) 2.8 ppb of pesticide; (F) return to the substrate.

The inhibition curves were constructed plotting the inhibition coefficient calculated according to Eq. (3), as a function of the pesticide concentration. In all the cases logarithmic equations were demonstrated to fit the experimental data. Limit of detection (LOD) was inferred by applying the Hubaux Vos method [9] in the linearity range. A comparison between biosensor with enzyme immobilized in a crosslinking matrix without and with nanostructured gold surface on electrodes was performed from the calibration curves. As reported in Fig. 4, for nanostructured Pt NEE SPS

**Fig. 3** Measurements with Dichlorvos with nanostructured biosensor, applying the standard addiction method in a steady state mode; A = only substrate; B = 0.5 ppb of Dichlorvos; C = 1 ppb; D = 1.8 ppb; E = 2.8 ppb; F = substrate in PB solution

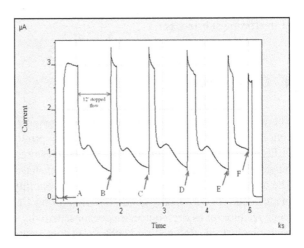

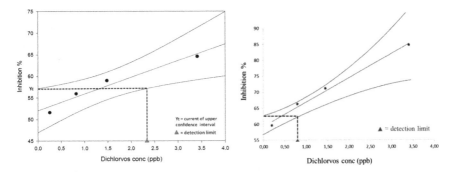

**Fig. 4** Graphical method to determine the LOD for Dichlorvos using the linear part of calibration [9]; graph on left for graphite AcChE bioelectrode, on right for the Pt NEEs nano bioelectrode

biosensors (sensor 6 of Table 1) a LOD for Dichlorvos of 0.8 $\mu g\,L^{-1}$ was extrapolated (right side): this value is three time lower than those obtained with sensor 2 of Table 1 (2.4 $\mu g\,L^{-1}$, left side).

Accuracy of the whole analytical method can be partially evaluated by fortifying fruit samples with a known amount of pesticides. Dichlorvos was spiked at a concentration of 12.5 ppb respectively in an organic strawberry sample at 1/150 dilution, and 12.5 ppb in an organic banana sample at 1/12 dilution, with buffer and substrate at unvaried experimental conditions, before each pretreatment. The recovery was calculated as in the relation: $\mathbf{R\% = (C_1 - C_0)/C_a * 100}$, where $C_1$ is the matrix concentration with added spike sample calculated from the calibration curve, $C_0$ is the concentration of pure matrix and $C_a$ is the theoretical added amount of pesticide.

The procedure was applied on three samples replicates comparing AcChE sensor nr. 2 with AcChE sensor nr. 6 of Table 1. For 12.5 ppb Dichlorvos added in strawberry sample, recovery percentages R% = 66.2 ± 5.8% and R% = 77.8 ± 8.1% were obtained, whereas added in banana sample R% = 101.8 ± 3.8 and R% = 104.0 ± 6.1%, respectively for screen printed sensor and NEE SPS sensor. Better recovery coefficients were obtained with banana samples, probably because of a lower content of interfering substances.

# 4   Conclusions

The presented work has been focused on enzymatic (bio)sensing hybrid systems, coupling nanostructured surfaces and screen printed technology, to detect OPs diffused in agricultural applications as cholinesterase inhibitors. Different immobilization techniques and active electrode surface have been used for such a purpose. The analytical performances confirmed the possibility to identify OPs at low levels of concentration, especially under flow conditions. Calibration have been performed using Dichlorvos as a reference. Biosensor sensitivity significantly higher in case of a nano-bioelectrode (0.8 ppb for Dichlorvos) than equivalent macroelectrode has been measured. Organic fruit samples have been spiked with known amount of contaminant, in order to measure and compare the recovery percentages. These results are a stimulating challenge to improve Au and Pt nano-bioelectrodes technology for OPs detection. Further trails and inter-laboratories measurements in simplified matrices are needed on certified reference materials (CRM) for a complete evaluation of the method accuracy.

# References

1. Krishnamoorthy K, Zoski CG (2005) Anal. Chem. 77:5068–5071 and ref. therein
2. Lu Q et al (2004) Nano Lett 4/12, 2473 and ref. therein
3. Delvaux M et al (2005) Biosens Bioelectronics 20:1587–1594 and ref. therein

4. Menon VP, Martin CR (1995) Anal Chem 67/13:1920–1928 and ref. therein
5. Vastarella W et al (2007) Intern J Environ Anal Chem 87/10–11, 701–714
6. Vastarella W et al (2009) Biosensing using Nanomaterial- chap. 13; ed. Merkoci, J. Wiley and Sons Inc. Publ., pp 379–419
7. Andreescu S, Marty JL (2006) Biomol. Engineering 23:1–15
8. Hildebrandt et al (2008) Sens Actuators B 133:195–201
9. Hubaux G (1970) Vos: Anal Chem 42/8:849

# Studies on $O_2$-$TiO_2$ Interplay Toward Unconventional MOX-Based Optodes for Oxygen Detection

**S. Lettieri, S. Amoruso, P. Maddalena, M. Alfè, V. Gargiulo, A. Fioravanti, and M. C. Carotta**

**Abstract** Titanium dioxide ($TiO_2$) is a functional semiconductor metal oxide that plays a major role in the many applicative fields, mostly including water remediation, surface functionalization and photoanodes for photoelectrochemical reactors. $TiO_2$ is also one of the known chemoresistive materials that can be adopted as sensitive layers in solid state gas sensors. $TiO_2$ has a peculiarly property in the fact that it exhibits two different photoluminescence (PL) spectra, depending on the crystalline polymorph considered (rutile or anatase), and that the PL emission of these two phases reacts differently as exposed to molecular oxygen ($O_2$). We show some representative results of this phenomenon, discussing the possibility to exploit it actual applications.

**Keywords** Photoluminescence · $TiO_2$ · Anatase · Rutile · Oxygen

## 1 Introduction

Quantitative detection of molecular oxygen ($O_2$) in gaseous and/or aqueous media is of great importance for many industrial, environmental, and biological applications. For example, the pollution levels in wastewater can be estimated via measurement of $O_2$ consumption by aerobic microorganisms, while the dissolved oxygen

S. Lettieri (✉)
Institute of Applied Sciences and Intelligent Systems (CNR-ISASI), Via Campi Flegrei 34, 80078 Pozzuoli, Italy
e-mail: stefano.lettieri@isasi.cnr.it

S. Amoruso · P. Maddalena
Physics Department "E. Pancini", University of Napoli "Federico II", Via Cintia 21, 80126 Napoli, Italy

M. Alfè · V. Gargiulo
Institute for Research On Combustion, National Research Council (CNR-IRC), Piazzale V. Tecchio 80, 80125 Napoli, Italy

A. Fioravanti · M. C. Carotta
Institute for Agricultural and Earthmoving Machines, National Research Council (CNR-IMAMOTER), Via Canal Bianco 28, Ferrara, Italy

G. Di Francia and C. Di Natale (eds.), *Sensors and Microsystems*,
Lecture Notes in Electrical Engineering 753,
https://doi.org/10.1007/978-3-030-69551-4_28

(DO) concentrations give important data on cellular functions (cell growth rate, metabolism, protein synthesis) [1].

DO monitoring is frequently performed by means of *optodes*, i.e. optical sensors based on the measurement of photoluminescence (PL) changes (in intensity and/or lifetime) in a luminescent material interacting with $O_2$. The method benefits from peculiar advantages of opto-chemical sensing approaches (e.g., electromagnetic immunity, electrical isolation, and possibility to engineer compact devices). Moreover, it works at room temperature and thus allows to avoid electrical circuitry and heaters (which are instead used by chemoresistive sensors).

Most optical sensors are based on quenchable luminescence of organic molecules. However, PL of metal oxides such as $TiO_2$, [2, 3] ZnO [4, 5] or $SnO_2$ [6, 7] has also been proposed for gas sensing. Recent literature indicates that the PL of titanium dioxide ($TiO_2$) nanoparticles can exhibit large sensitivities to $O_2$ exposure [8]. In addition, it shows a peculiar feature which has no known equivalent among other metal oxides: the two stable $TiO_2$ polymorphs (anatase and rutile) have different photoluminescence spectra [9] with opposite changes in in the PL intensity vs. $O_2$ exposure, a special feature that suggest possibilities to enhance the sensitivity of a $TiO_2$-based optodes [10].

In this contribution, we review some of the most important results involving the $TiO_2/O_2$ interaction effect on the near-infrared and visible photoluminescence of $TiO_2$ nanostructures, discussing them in the framework of possible applications in the field of $O_2$ optodes.

## 2   Experimental Results

### 2.1   Preparation of TiO$_2$ Nanoparticles by Ultrafast Laser Ablation

Nanoparticulated $TiO_2$ films were produced via ultrafast laser ablation, employing femtosecond (fs) laser pulses of a frequency-doubled Neodymium-doped yttrium lithium fluoride (Nd:YLF) laser, providing laser pulses of ~300 fs time duration (full width at half maximum) at 527 nm wavelength. The laser beam was focused at 45∘ onto the a commercial $TiO_2$ rutile target using a 25 cm focal length lens. Typical spot area on target of about $5 \times 10^{-4}$ cm$^2$ were employed, with target ablation produced at laser fluence of about 1.4 J/cm$^2$.

Different background oxygen pressures ($p$) affect the kinetic energy of the $TiO_2$ aggregates ejected from the substrate and moving toward the substrate located 4.0 cm apart from the target. Morphological analysis of the nanoparticulated films formed on a silicon substrate allowed to characterize different pressure regimes, finding out that for $p \sim 1$ mbar the laser ablation led to assemblies of nanoparticles (NPs), with typical diameters in the 20–200 nm range.

The use of lower pressures (p ~ $10^{-3}$–$10^{-2}$ mbar) led to increased deposition rate but also to partial NPs coalescence, caused by the larger kinetic energy of the aggregates. As a result, the corresponding deposited films exhibited large globular structures and hence to lower effective surface area.

Other measurements were performed on commercial anatase NPs (Sigma Aldrich) and on Degussa P25 $TiO_2$ (mixed phase anatase/rutile nanoparticles). The latter was used to produce pure rutile powders by annealing at 800 °C for 1 h.

## 2.2 Rutile TiO₂ Photoluminescence

PL emission of rutile $TiO_2$ lies in the near-infrared ("NIR-PL") range, being typically peaked at a wavelength of about 840 nm (1.47 eV photon energy). The origin of such emission band is debated and, to this day, not completely understood. Some works underlined a close correlation between the NIR-PL emission band of rutile and the mechanism of water photooxidation under anodic bias, arguing that the emitting state is an intermediate of the reactions leading to $O_2$ photo-evolution and attributing the NIR PL to the radiative recombination of free electrons with self-trapped holes [11]. On an opposite side, other works attributed the NIR-PL to recombination of deeply-trapped electrons (located in the bulk of the nanocrystals) with holes localized close to the surface of the nanocrystal due to upward band bending caused by $O_2$ adsorption [12].

A representative example or NIR-PL enhancement caused by $O_2$ adsorption on rutile titania is reported in Fig. 1 for pure rutile films prepared by annealing of commercial P25 powders. The data correspond to $O_2$ concentrations of 0.5% and 0.25% in a $N_2$ + $O_2$ mixture obtained by flowing certified nitrogen and air bottles (300 sccm total flow).

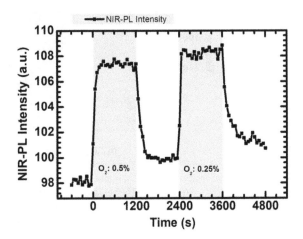

**Fig. 1** NIR-PL intensity of a rutile $TiO_2$ film measured during the exposure to different concentrations of $O_2$. The $O_2$ steps at concentrations 0.5% and 0.25% are followed by recovery in pure nitrogen flow. All the measurements are made at room temperature

## 2.3  *Anatase TiO₂ Photoluminescence*

PL emission of anatase TiO$_2$ occurs in the visible range, typically extending from
about 450 nm up to 700 nm. When excited by supra-gap light, such a PL emission
is quenched by exposure to O$_2$ over the entire emission range, although the intensity
modulations are more evident for the light emitted in the green range. A representative
example is shown in Fig. 2 for anatase films prepared by PLD. In the top graph it
is reported the total Pl intensity integrated over two spectral ranges, namely in the
green region (530–600 nm) and in the red region (680–750 nm). The composition
of air/nitrogen mixtures flown in the test chamber was varied over the values of 2%,
5%, 10% and 20% (relative percentage of air with respect to the total gas flow),

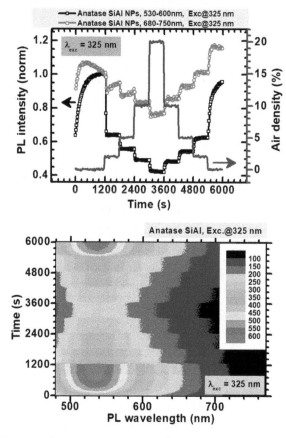

**Fig. 2** Top: PL intensity of anatase TiO$_2$ integrated in the wavelength intervals 530–600 nm (black
curve) and 680–750 nm (red curve). The blue curve indicates the air concentration in the air/nitrogen
mixtures. Bottom: image of the PL spectra intensity vs time during the experiment, evidencing the
reversible quenching of anatase PL in the green region. The excitation was provided by a Xe lamp
spectrally filtered at 325 nm through a double-grating monochromator

thus corresponding to oxygen concentrations of $[O_2] = 0.4\%$, 1%, 2% and 4%. The response is sublinear versus $[O_2]$, indicating that a Langmuir-type adsorption with a saturation onset lesser than 0.4% concentration. More interestingly, the response to $O_2$ is quite reversible once an optimal emission wavelength range is chosen (here the green emission, while the red emission shows larger amounts of fluctuations and instability).

The results suggest that the physical mechanisms that determine the Pl emission in the red and in the green region are different. This topic, partially debated in other works [13], is not discussed here.

# 3 Conclusions

Rutile and anatase $TiO_2$ PL experience a peculiar "anti-correlated" intensity response upon $O_2$ exposure. This characteristic, which to the best of our knowledge is has no analogues in other metal oxides, allows in principle to implement an increase in optical sensitivity to $O_2$ by ratiometric detection in mixed-phase $TiO_2$ systems. Such an approach would exhibit favorable properties, mainly due to the possibility to tackling the need to develop supporting matrixes that carry the luminescent probes and to avoid problems caused by the use of non-homogeneous components in the optical ratiometric sensor.

# References

1. Quaranta M, Borisov SM, Klimant I (2012) Indicators for optical oxygen sensors. Bioanalytical Rev 4(2–4):115–157. https://doi.org/10.1007/s12566-012-0032-y
2. Setaro A, Lettieri S, Diamare D, Maddalena P, Malagù C, Carotta MC, Martinelli G (2008) Nanograined anatase titania-based optochemical gas detection. New J Phys 10(5):053030. https://doi.org/10.1088/1367-2630/10/5/053030
3. Pallotti D, Orabona E, Amoruso S, Maddalena P, Lettieri S (2014) Modulation of mixed-phase titania photoluminescence by oxygen adsorption. Appl Phys Lett 105(3):031903. https://doi.org/10.1063/1.4891038
4. Cretì A, Valerini D, Taurino A, Quaranta F, Lomascolo M, Rella R (2012) Photoluminescence quenching processes by $NO_2$ adsorption in ZnO nanostructured films. J Appl Phys 111(7):073520. https://doi.org/10.1063/1.3700251
5. Carotta MC, Cervi A, Gherardi S, Guidi V, Malagu' C, Martinelli G, Vendemiati B, Sacerdoti M, Ghiotti G, Morandi S, Lettieri S, Maddalena P, Setaro A (2009) (Ti, Sn)$O_2$ Solid solutions for gas sensing: a systematic approach by different techniques for different calcination temperature and molar composition. Sensors Actuators B Chem 139(2):329–339. https://doi.org/10.1016/j.snb.2009.03.025
6. Prades J, Arbiol J, Cirera A, Morante J, Avella M, Zanotti L, Comini E, Faglia G, Sberveglieri G (2007) Defect study of $SnO_2$ nanostructures by cathodoluminescence analysis: application to nanowires. Sensors Actuators B Chem 126(1):6–12. https://doi.org/10.1016/j.snb.2006.10.014
7. Trani F, Causà M, Lettieri S, Setaro A, Ninno D, Barone V, Maddalena P (2009) Role of surface oxygen vacancies in photoluminescence of Tin dioxide nanobelts. Microelectron J 40(2):236–238. https://doi.org/10.1016/j.mejo.2008.07.060

8. Pallotti DK, Passoni L, Gesuele F, Maddalena P, Di Fonzo F, Lettieri S (2017) Giant $O_2$-induced photoluminescence modulation in hierarchical Titanium Dioxide nanostructures. ACS Sensors 2(1):61–68. https://doi.org/10.1021/acssensors.6b00432

9. Pallotti DK, Orabona E, Amoruso S, Aruta C, Bruzzese R, Chiarella F, Tuzi S, Maddalena P, Lettieri S (2013) Multi-band photoluminescence in $TiO_2$ nanoparticles-assembled films produced by femtosecond pulsed laser deposition. J Appl Phys 114(4):043503. https://doi.org/10.1063/1.4816251

10. Lettieri S, Pallotti DK, Gesuele F, Maddalena P (2016) Unconventional ratiometric-enhanced optical sensing of oxygen by mixed-phase $TiO_2$. Appl Phys Lett 109(3):031905. https://doi.org/10.1063/1.4959263

11. Imanishi A, Okamura T, Ohashi N, Nakamura R, Nakato Y (2007) Mechanism of water photooxidation reaction at atomically flat $TiO_2$ (Rutile) (110) and (100) surfaces: dependence on solution PH. J Am Chem Soc 10

12. Vequizo JJM, Kamimura S, Ohno T, Yamakata A (2018) Oxygen induced enhancement of NIR emission in Brookite $TiO_2$ powders: comparison with Rutile and Anatase $TiO_2$ powders. Phys Chem Chem Phys 20(5):3241–3248. https://doi.org/10.1039/C7CP06975H

13. Rex RE, Knorr FJ, McHale JL (2014) Surface traps of $TiO_2$ nanosheets and nanoparticles as illuminated by spectroelectrochemical photoluminescence. J Phys Chem C 118(30):16831–16841. https://doi.org/10.1021/jp500273q

# A Preliminary Characterization of an Air Contaminant Detection System Based on a Multi-sensor Microsystem

**L. Gerevini, C. Bourelly, G. Manfredini, A. Ria, B. Alfano, S. De Vito, E. Massera, M.L. Miglietta, and T. Polichetti**

**Abstract** The air quality is a fundamental aspect for human health and its monitoring is a crucial task that needs to be performed. In this framework, an IoT ready proposal is here reported to perform both the sensing and the classification of possible air contaminants. In particular, the paper focuses on the analysis of the measurement data, by characterizing them from a metrological point of view. In detail, after having designed a suitable measurement procedure by adopting several sensors available on the platform, data are acquired during clean air exposition, air contaminant injection and steady state conditions, where contaminant has been fully injected and measure-

L. Gerevini (✉)
Department of Electrical and Information Engineering, University of Cassino and Southern Lazio, Via G. Di Biasio, 43, 03043 Cassino, Italy
e-mail: luca.gerevini@unicas.it

C. Bourelly
Sensichips s.r.l., Via delle Valli, Aprilia, Italy
e-mail: carmine.bourelly@sensichips.com

G. Manfredini · A. Ria
Department of Information Engineering, University of Pisa, Via Caruso 16, Pisa, Italy
e-mail: giuseppe.manfredini@phd.unipi.it

A. Ria
e-mail: andrea.ria@ing.unipi.it

B. Alfano · S. D. Vito · E. Massera · M.L. Miglietta · T. Polichetti
ENEA, Piazzale Enrico Fermi, 1, Portici, Italy
e-mail: brigida.alfano@enea.it

S. D. Vito
e-mail: saverio.devito@enea.it

E. Massera
e-mail: ettore.massera@enea.it

M.L. Miglietta
e-mail: mara.miglietta@enea.it

T. Polichetti
e-mail: tiziana.polichetti@enea.it

© The Author(s), under exclusive license to Springer Nature Switzerland AG 2021
G. Di Francia and C. Di Natale (eds.), *Sensors and Microsystems*,
Lecture Notes in Electrical Engineering 753,
https://doi.org/10.1007/978-3-030-69551-4_29

ments have reached a stable state. Considerations arising from the analysis of acquired data are the basis for a classification stage, based on machine learning mechanisms. Definitively, this work represents an extension of previously presented results, and it shows how the sensor technology has been greatly improved, thus allowing to feed the detection systems with higher reliable results and therefore enhancing their performance.

**Keywords** Contaminant detection · Air monitoring · Sensor networks · IoT

# 1   Introduction

Air quality in indoor and outdoor environments is a primary development factor for human civilization. In detail, several threats impact on air pollution conditions and a reliable monitoring system is crucially needed especially in city environments, where industrial and car emission effects deeply influence the air quality levels [3, 16, 17]. Currently, monitoring actions are managed through costly and bulky devices where professional technicians are needed to perform measurement tasks. On the contrary, low-cost solutions arise from the academic and on-the-edge industrial fields. Such solutions are sensor-based and need an important effort in terms of proper sensor choice, conditioning and data processing but they warrant low-cost, low-power and high flexibility features [15]. As widely recognized in literature, the implementation of a sensor-based solution often requires the adoption of sensor arrays, in order to exploit their different sensitivity to contaminant materials dispersed in air [9, 13]. In this sense, the authors have developed a solution based on a proprietary multi-sensing platform, endowed with suitable acquisition, conditioning, data transmission and processing mechanisms to ensure an accurate classification system aided by Artificial Intelligence (AI). A very important role is played by the measurement mechanism, the quality of the raw data and the different sensor's sensitivity to different substances. For this reason, stemming from the authors' experience in data analysis, sensor fusion, classification systems [1, 2, 4, 5, 7, 8, 10–12, 14], here we propose a detailed metrological characterization of the sensing activity performed through the platform itself. The characterization has been made by analysing crucial properties for the adopted sensors, as their stability in nominal conditions (exposed to clean air), their reactivity to contaminant injection and their steady-state behavior when they are exposed to contaminant for prolonged times. In the paper, a comparison with a previous release of the adopted platform is also provided, thus highlighting how the on-going work is leading towards miniaturization, sensor integration and higher sensitivity to air contaminants. In detail, for each substance, several sensors are adopted and electrical parameters (resistance and capacitance) are measured at different frequencies. For each stable configuration, ten trials are performed by controlling the environment and contaminant concentrations conditions.

## 2 Sensichips Smart Cable Air Board

The Smart Cable Air (SCA) board is developed by Sensichips s.r.l. with the aim to create a compact size, about 8×9 mm, and low-power solution to monitor the air quality. The board includes a sensor array composed of: one aluminum oxide sensor on SENSIPLUS chip, two polymeric sensors, one on SENSIPLUS chip and the other one on the containing board, a CMOS bandgap temperature sensor and a MEMS gas sensor (FIGARO tgs8100). More details about the on-board sensors can be found in [14]. The polymeric sensor on the board is made with polyaniline nanofibers and synthetized according to experimental procedure reported in [6]. Briefly, the initiator solution (ammonium peroxydisulfate in camphorsulfonic acid CSA buffered water) were mixed into the monomer solution (aniline in CSA buffered water). A rapid polymerizing aniline monomers induce the formation of short fibres with a diameter of the order of 50–100 nm and a length from hundreds of nanometres to several micrometers. SENSIPLUS is a proprietary technology of Sensichips designed in collaboration with the University of Pisa.

## 3 Measurements

### 3.1 Measurements Setup

To achieve good classifier performance, a data-set is needed. Measurement setup and protocol are fundamental. A cubic plexiglass chemical hood 300 × 300 mm is created (see Fig. 1), with two fans to extract polluted air. Inside the hood, there are three SCAs allowing to have, at the same time, three different acquisitions under the same conditions. The chips are placed face down at 4 cm from the point of injection of the analyte.

The SCAs are connected to the PC via three different MCUs, acting as a communication bridge between the SENSIPLUS chips and PC. A batch of eight different pollutants is used to create the dataset.

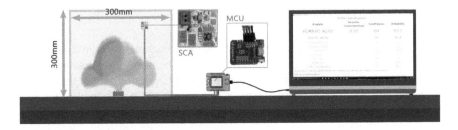

**Fig. 1** Measurement Setup

The pollutants have been injected via a syringe into a little plate placed under the three chips inside the chemical hood. The list of all substances and quantity, used for the experiments, are listed in the Table 1.

Any experiment consist in the acquisition of 1600 samples that are around 50 min of measurement at 23°C. Since it was not possible to measure the amount of evaporated substance, it is difficult to estimate the ppm of a given pollutant.

## 3.2   Measurement Protocol and Dataset Structure

In order to collect a big data-set on which we can study and evaluate the improvement of our sensor technology performance, many data have been collected.
Measurements protocol consists of experiments that contain a 1600 samples, structured as:

- warm-up acquisition: the first 300 samples have been taken in clean air in order to build a good reference for the measurement. The warm-up phase gives a starting point to evaluate the sensitivity and its value can be also used for sensor calibration;
- substance acquisition: 1000 samples taken in pollutant presence;
- sensor recovery: 300 samples are acquired meanwhile the substance is removed from the environment in order to evaluate the sensors' recovery capability.

The data-set is based on 10 acquisitions, as described above, and the process is repeated for all different tested pollutants.

## 4   Improvement Evaluation

In this paper we show a comparison of the new release of our SCA platform (Version 2) with the previous one (Version 1). For the sake of the evaluation of the improvement, we made a study about the response ($S_r$) computed for each sensor of the array over all the acquisitions by considering:

$$S_r = \frac{|M_{steady} - M_{warm-up}|}{\sigma_{warm-up}} \tag{1}$$

where the $M_{steady}$ has been taken as mean value on the samples measured in the substance acquisition phase, whereas the $M_{warm-up}$ on the warm-up acquisition phase samples (see Sect. 3.2) and the $\sigma_{warm-up}$ is the standard deviation of each sensor. This parameter tells us if the response of a given sensor is due to the reaction with a contaminant or it is due to some measurement effects like noise, drift, etc. Furthermore, in order to evaluate the improvement of our system, we computed the coefficient of variation $\sigma_{Sens}^{\star}$, that is a standardized measure of dispersion of a probability distribution, associated to each feature, over all the acquisitions:

**Table 1** List of Substances

| Substance | Ethanol | Ammonia | Acetone | Nitromethane | Water | Acetic Acid | Formic Acid | Hydrocloric Acid |
|---|---|---|---|---|---|---|---|---|
| Quantity [ml] | 1.00 | 1.80 | 1.10 | 0.56 | 0.54 | 1.40 | 1.30 | 5.00 |

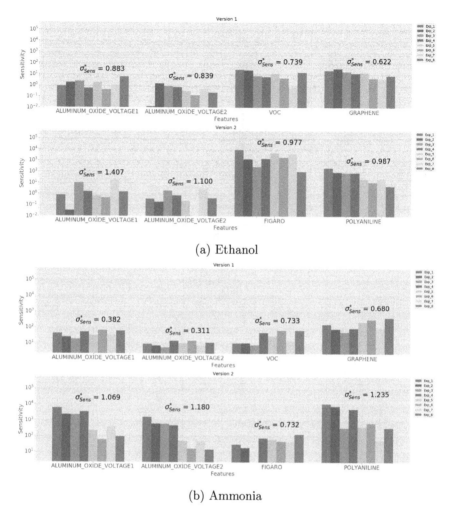

(a) Ethanol

(b) Ammonia

**Fig. 2** Sensitivity and coefficient of variation comparison results with two different substances

$$\sigma_{Sens}^{\star} = \frac{\sigma_{Sens,Acq}}{|\mu_{Sens,Acq}|} \tag{2}$$

where $|\mu_{Sens,Acq}|$ and $\sigma_{Sens,Acq}$ are the mean value and the standard deviation for each sensor respectively. Figure 2a shows the comparison results between the SCA Version 1 and 2 with the ethanol.

Even though we cannot perform a one by one comparison due to the different nature of sensors, we can see an improvement on the sensors response in the SCA Version 2. In fact, the sensor's response of all the sensors is quite similar except to the FIGARO ones that show a response about two orders of magnitude greater. Moreover, another improvement resides in the miniaturization of some sensors that allow the SCA version 2 to integrate a greater number of sensors in order to better exploit the different sensitivity to contaminants.

Another results are reported in Fig. 2b, where we show the comparison between the two versions of SCA exposed to the ammonia. As it can be seen in the image, in Version 2 the two *ALUMINUM_OXIDE* sensors reached a sensitivity value of about two orders of magnitude greater then the ones reached by Version 1. In this case it is possible to see an improvement caused by the technology evolution of the SCA. Of course more consideration can be done with different substances but they cannot be reported due space limitation issues.

# 5   Conclusion

In this work, we've reported a feature comparison between the new SCA with the old one. In order to fairly compare the two systems, we used two figure of merit: *sensor response* and *coefficient of variation* as described in the previous sections. From the obtained results, thus, we can assert that the miniaturization of some sensors (that allow us to add some new sensors to the SCA as FIGARO) took us to an overall improvement of the entire systems in terms of *sensor's response*.

**Acknowledgements** The research leading to these results has received funding from the European Union's Horizon 2020 research and innovation programme under grant agreement SYSTEM No. 787128. The authors are solely responsible for it and that it does not represent the opinion of the Community and that the Community is not responsible for any use that might be made of information contained therein.

This work was also supported by MIUR (Minister for Education, University and Research, Law 232/216, Department of Excellence).

# References

1. Angrisani L, Capriglione D, Cerro G, Ferrigno L, Miele G (2016) On employing a Savitzky-golay filtering stage to improve performance of spectrum sensing in cr applications concerning vdsa approach. Metrol Measurement Syst 23(2):295–308. https://doi.org/10.1515/mms-2016-0019

2. Angrisani L, Capriglione D, Cerro G, Ferrigno L, Miele G (2016) Optimization and experimental characterization of novel measurement methods for wide-band spectrum sensing in cognitive radio applications. Measurement 94:585–601. https://doi.org/10.1016/j.measurement.2016.08.036. http://www.sciencedirect.com/science/article/pii/S0263224116304973

3. Bentayeb M, Wagner V, Stempfelet M, Zins M, Goldberg M, Pascal M, Larrieu S, Beaudeau P, Cassadou S, Eilstein D, Filleul L, Le Tertre A, Medina S, Pascal L, Prouvost H, Quénel P, Zeghnoun A, Lefranc A (2015) Association between long-term exposure to air pollution and mortality in france: a 25-year follow-up study. Environ Int 85. https://doi.org/10.1016/j.envint.2015.08.006

4. Bernieri A, Ferrigno L, Laracca M, Rasile A (2017) An AMR-based three-phase current sensor for smart grid applications. IEEE Sensors J 17(23):7704–7712

5. Betta G, Cerro G, Ferdinandi M, Ferrigno L, Molinara M (2019) Contaminants detection and classification through a customized iot-based platform: a case study. IEEE Instrumentation Measurement Mag 22(6):35–44

6. Borriello A, Guarino V, Schiavo L, Alvarez-Perez MA, Ambrosio L (2011) Optimizing pani doped electroactive substrates as patches for the regeneration of cardiac muscle. J Mater Sci Mater Med 22:1053–1062. https://doi.org/10.1007/s10856-011-4259-x

7. Bria A, Cerro G, Ferdinandi M, Marrocco C, Molinara M (2020) An iot-ready solution for automated recognition of water contaminants. Pattern Recogn Lett 135:188–195. https://doi.org/10.1016/j.patrec.2020.04.019

8. Bruschi P, Cerro G, Colace L, De Iacovo A, Del Cesta S, Ferdinandi M, Ferrigno L, Molinara M, Ria A, Simmarano R, Tortorella F, Venettacci C (2018) A novel integrated smart system for indoor air monitoring and gas recognition. In: 2018 IEEE International Conference on Smart Computing (SMARTCOMP), pp 470–475

9. Castell N, Dauge F, Schneider P, Vogt M, Lerner U, Fishbain B, Broday D, Bartonova A (2016) Can commercial low-cost sensor platforms contribute to air quality monitoring and exposure estimates? Environ Int 99. https://doi.org/10.1016/j.envint.2016.12.007

10. Cerro G, Ferdinandi M, Ferrigno L, Laracca M, Molinara M (2018) Metrological characterization of a novel microsensor platform for activated carbon filters monitoring. IEEE Trans Instrumentation Measurement 67(10):2504–2515

11. Cerro G, Ferdinandi M, Ferrigno L, Molinara M (2017) Preliminary realization of a monitoring system of activated carbon filter rli based on the sensiplus® microsensor platform. In: 2017 IEEE international workshop on Measurement and Networking (M N), pp 1–5

12. Ferdinandi M, Molinara M, Cerro G, Ferrigno L, Marrocco C, Bria A, Di Meo P, Bourelly C, Simmarano, R.: A novel smart system for contaminants detection and recognition in water. In: 2019 IEEE international conference on Smart Computing (SMARTCOMP), pp 186–191 (2019)

13. Li Y, Pang W, Sun C, Zhou Q, Lin Z, Chang Y, Li Q, Zhang M, Duan X (2019) Smartphone-enabled aerosol particle analysis device. IEEE Access 7:101117–101124

14. Molinara M, Ferdinandi M, Cerro G, Ferrigno L, Massera E (2020) An end to end indoor air monitoring system based on machine learning and sensiplus platform. IEEE Access 8:72204–72215. https://doi.org/10.1109/ACCESS.2020.2987756

15. Morawska L et al (2018) Applications of low-cost sensing technologies for air quality monitoring and exposure assessment: how far have they gone? Environ Int 116:286–299. https://doi.org/10.1016/j.envint.2018.04.018

16. Santos C, Jiménez JA, Espinosa F (2019) Effect of event-based sensing on IoT node power efficiency. case study: air quality monitoring in smart cities. IEEE Access 7:132577–132586 (2019)

17. World Health Organization (2016) Ambient air pollution: a global assessment of exposure and burden of disease. World Health Organization

# A Preliminary Characterization of a Water Contaminant Detection System Based on a Multi-sensor Microsystem

C. Bourelly, L. Gerevini, M. Cicalini, and G. Manfredini

**Abstract** The evaluation of measurement quality is a primary task when performing sensing activities based on low–cost sensor technologies. In detail, in this work electrical impedance measurements on different sensors are proposed in order to have a dataset of raw data to be used to perform classification of possible contaminants in water environment. In detail, the sensor technology is based on a proprietary multi-sensing platform which is arranged in a suitable set-up able to carry out measurements in water for prolonged times. Sensors metalized with different materials are jointly used to exploit sensitivity diversity to different contaminants. An ad–hoc measurement procedure has been designed, including data acquisition during warm–up period, contaminant injection and steady state conditions. Since different releases of the platform have been developed, here we propose a comparison between two versions to demonstrate the technological advance in sensor integration and miniaturization leading to higher reliable results. A metrological analysis of the obtained measurements with two different platform versions is carried out and compared results are reported.

**Keywords** Contaminant detection · Water monitoring · Sensor networks · Machine learning · Iot

C. Bourelly
Sensichips S.r.l., Aprilia, Italy
e-mail: carmine.bourelly@sensichips.com

L. Gerevini (✉)
Department of Electrical and Information Engineering, University of Cassino
and Southern Lazio, Cassino, Italy
e-mail: luca.gerevini@unicas.it

M. Cicalini · G. Manfredini
Department of Information Engineering, University of Pisa, Via Caruso 16, Pisa, Italy
e-mail: mattia.cicalini@phd.unipi.it

G. Manfredini
e-mail: giuseppe.manfredini@phd.unipi.it

© The Author(s), under exclusive license to Springer Nature Switzerland AG 2021
G. Di Francia and C. Di Natale (eds.), *Sensors and Microsystems*,
Lecture Notes in Electrical Engineering 753,
https://doi.org/10.1007/978-3-030-69551-4_30

223

# 1 Introduction

Contaminants' detection and classification are fundamental tasks to monitor water status and to control its pollution levels, since several negative effects, as the increase in human death rate [1], are provoked by water poor cleanliness. Human actions often threat environments and especially water conditions, due to industrial and agricultural intensive activities. Marine, freshwater, tap drinkable water are all victims of pollution [2]. It is therefore crucial to have reliable technologies able to measure, process and classify water–related data in a very fast way to ensure minimization of damages deriving from the usage of polluted water for domestic purposes. Such activities must be performed in form of continuous monitoring, to obtain a water conditions' map, both in spatial and temporal dimensions [3, 4]. The state of the art presents also complex systems, able to do that with high accuracy but also high cost and often poorly portable solutions. On the other hand, facing challenging problems and working with ad–hoc solutions, it is possible to deal with it by means of sensor technologies, currently envisioned and considered as the most promising way to face the issue in a controlled, low–cost, low–power, highly accurate manner [5]. Actually, they also open several issues: which sensors have to be used, how data must be acquired, how to process big amount of raw measurements and what are the quality factors raw data must have to ensure good classification capabilities. For sure, machine learning is widely recognized as suitable tool to manage raw data and to properly detect the presence of contaminants and classify them. In this sense, the authors have already experienced it in [6–8]. Nevertheless, the quality of the raw measurement is a primary goal to achieve: choose the most sensitive sensors, characterize them from the stability point of view, carry out reliable measurement campaign, by choosing the most reliable parameters to monitor (electrical, chemical, environmental). Taking into account such premises, stemming on our past experience on sensors and data processing [9–12], here we present a metrological analysis of a proprietary multi-sensing platform developed by Sensichips s.r.l. [13] and the Department of Information Engineering of the University of Pisa, by providing its capabilities to measure electrical properties and combining multi–sensor information to obtain high sensitivity to different kinds of pollutants. In particular, in [14], we already provided some insights about contaminant classification in water by the same proprietary platform and here we describe the on–going progress, by demonstrating how a greater sensor integration and technological improvement has led us to improve sensitivity and repeatability of the measurement process, thus preparing the system to have better classification results.

# 2 Sensichips' Smart Cable Water Board

In its latest release, Sensichips developed the Smart Cable Water (SCW) with the intention to create a contaminants' detection board, based on impedance spectroscopy measurements over six different coated interdigitated electrodes (IDEs): five small IDEs on the front face coated with Gold, Copper, Silver, Nickel, Palladium (dimensions 3 mm by 7 mm each) and one Platinum IDE (dimensions 12 mm by 8 mm) on the backside. A CMOS bandgap temperature sensor, a light emitter and a light photo-diode sensor are also available on the board. Version one board has only two IDEs, made out of gold and copper, with dimensions 9 mm by 10 mm each. The core of the SCW boards is the SENSIPLUS chip, a proprietary technology of Sensichips developed in collaboration with the University of Pisa.

## 3 Measurements

### 3.1 Measurements Setup

All measurements must be acquired in the same reliable conditions that abate as much as possible all the interferents variables, clearly is not possible achieve measurements in sewage network, because the background environmental composition is not stable. Otherwise, use clean water could not be the best emulation of the destination environment. The solution adopted is emulate the sewage water and sewage network. A Synthetic Waste Water (SWW) is created, based on a simplified recipe in [15]. For the sewage moving water emulation a magnetic stirrer with a 25 mm anchor is used. A 300 ml beaker is placed on top and the stirring speed is set to 50 rpm. The beaker is filled with 100 ml of SWW, with the Smart Cable Water hooked on the beaker side. A PC equipped with Sensichips' "*Winux*" software control the Smart Cable Water through an MCU that only converts the proprietary Sensichips' bus to USB protocol. Figure 1 shows a setup depiction A batch of nine different pollutants is used to create the data-set.

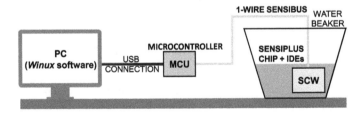

**Fig. 1** Measurement setup

## 3.2   Measurement Protocol

Sensichips' SCW is used to measure impedance at two different frequencies over four different sensing ports sequentially. Every measurement output has 10 features:

- Resistance at 78 kHz for platinum and gold interdigitated electrodes (IDEs),
- Resistance and Capacitance 200 Hz for Platinum, Gold, Silver, and Nickel IDEs.

Each measure is composed of a 300 only synthetic wastewater baseline samples, followed by 1000 samples after pollutant injection. The measurement quantities have been chosen according to the discriminant capabilities of the sensors: as an example, Copper IDE has resulted less sensitive to the contaminants under test and, therefore, Resistance and Capacitance values have been accordingly discarded. The resulting data-set is organized as follows. Each experiment is composed of ten physical quantities measured (depicted in ten columns): two resistance at 78 kHz, four resistance and four capacitance 200 Hz; by 1600 samples (depicted in rows): first 300 discarded, further 300 for baseline acquisition and last 1000 for substance response. A total of ten experiments per substance is taken.

## 4   Measurement Results

In order to show the improvement of the novel release (namely, version 2) with respect to what happened with version 1 [14], we provide here results using two figures of merit, customized to our case. The first one is the sensitivity: for each experiment, we compute the ratio between the total variation of the measured quantity due to contaminant injection and the standard deviation of the samples acquired during water–only presence. In other words, it states if the measurand variation is effectively due to the contaminant or it can be considered as a random fluctuation of baseline condition. Assuming a Gaussian model, we consider to have contaminant effect on the measurement when the sensitivity value exceeds 3.

$$Sensitivity = \frac{|M_{steady} - M_{baseline}|}{\sigma_{baseline}} \tag{1}$$

In Eq. 1, $M_{steady}$ and $M_{baseline}$ represent the mean measurement value during steady–state period (after contaminant injection) and baseline period (before contaminant injection), respectively.

The second figure of merit is the coefficient of variation applied to the Sensitivity, over the repeated experiments, for each feature.

$$\sigma_{Sens}^{\star} = \frac{\sigma_{Sens,Exp}}{|\mu_{Sens,Exp}|} \tag{2}$$

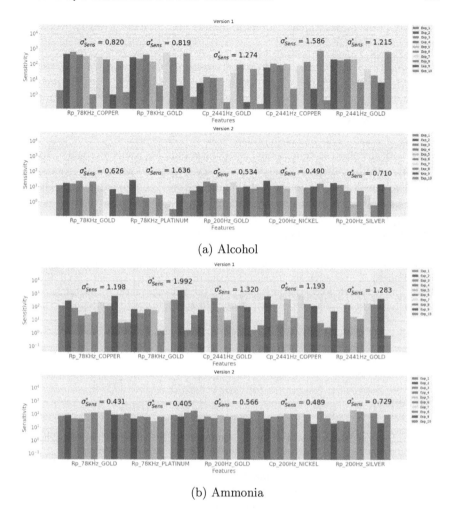

(a) Alcohol

(b) Ammonia

**Fig. 2** Sensitivity Comparison result with two different substances

Results are provided in Figs. 2a, b, for two specific substances, Alcohol and Ammonia. In such figures, performance is reported in terms of *Sensitivity* in form of histogram, for a subset of the measured quantities (features). Furthermore, in text mode, the *coefficient of variation* ($\sigma^{\star}_{Sens}$) is depicted in the upper part, one for each set of experiment and feature. Clearly, due to the technological evolution of the chip, even the features and the sensors have been changed. Nevertheless, what we want to prove is that, despite a very higher level of integration, resulting in a smaller exposure surface, the Sensitivity is substantially the same while the coefficient of variation is generally decreased. We have nearly the same capacity to react to contaminants, but we have more homogeneous results over repeated experiments, thus increasing the reliability of the measurements and more sensors to exploit the diversity in classification phase. Similar considerations can be drawn by considering other substances in the contaminant set.

## 5  Conclusion

In this work an enhanced contaminant detection system has been presented by focusing on its metrological features and comparing it with its previous version, from which a sensitive sensor integration and technological efforts voted to reduce the exposed surface has led to an improvement in terms of sensors' response (here named as sensitivity) and reduced dispersion (coefficient of variation). Such results allow the system to be tested in classification mode, by adopting machine learning facilities.

**Acknowledgements**  The research leading to these results has received funding from the European Unions Horizon 2020 research and innovation programme under grant agreement SYSTEM No. 787128. The authors are solely responsible for it and that it does not represent the opinion of the Community and that the Community is not responsible for any use that might be made of information contained therein

This work was also supported by MIUR (Minister for Education, University and Research, Law 232/216, Department of Excellence).

## References

1. Prüss-Üstün A, Wolf J, Corvalán C, Bos R, Neira M (2016) Preventing disease through healthy environments: a global assessment of the burden of disease from environmental risks. World Health Organization
2. Goel P (2006) Water Pollution: Causes. Effects and Control, New Age International
3. Hall J, Zaffiro AD, Marx RB, Kefauver PC, Krishnan ER, Haught RC, Herrmann JG (2007) On-line water quality parameters as indicators of distribution system contamination. J AWWA 99(1):66–77
4. Li J, Cao S (2015) A low-cost wireless water quality auto-monitoring system. Int J Online Eng (iJOE) 11(3):37–41
5. Jamison DT, Breman JG, Measham AR, Alleyne G, Claeson M, Evans DB, Jha P, Mills A, Musgrove P (2006) Disease control priorities in developing countries. The international bank for reconstruction and development/The World Bank 2006
6. Bruschi P, Cerro G, Colace L, De Iacovo A, Del Cesta S, Ferdinandi M, Ferrigno L, Molinara M, Ria A, Simmarano R, Tortorella F, Venettacci C (2018) A novel integrated smart system for indoor air monitoring and gas recognition. IEEE international conference on smart computing (SMARTCOMP), 2018, pp 470–475
7. Bria A, Cerro G, Ferdinandi M, Marrocco C, Molinara M (2020) An iot-ready solution for automated recognition of water contaminants. Pattern Recogn Lett 135:188–195 (2020). [Online]. Available: http://www.sciencedirect.com/science/article/pii/S0167865520301410
8. Molinara M, Ferdinandi M, Cerro G, Ferrigno L, Massera E (2020) An end to end indoor air monitoring system based on machine learning and sensiplus platform. IEEE Access 8:72 204–72 215

9. Cerro G, Ferdinandi M, Ferrigno L, Laracca M, Molinara M (2018) Metrological characterization of a novel microsensor platform for activated carbon filters monitoring. IEEE Trans Instrum Meas 67(10):2504–2515
10. Ferrigno L, Laracca M, Liguori C, Pietrosanto A (2012) An FPGA-based instrument for the estimation of $R$, $L$, and $C$ parameters under nonsinusoidal conditions. IEEE Trans. Instrum. Meas. 61(5):1503–1511
11. Angrisani L, Capriglione D, Cerro G, Ferrigno L, Miele G (2016) Optimization and experimental characterization of novel measurement methods for wide-band spectrum sensing in cognitive radio applications. Measurement 94:585–601
12. Angrisani L, Capriglione D, Cerro G, Ferrigno L, Miele G (2016) On employing a savitzky-golay filtering stage to improve performance of spectrum sensing in CR applications concerning VDSA approach. Metrol Meas Syst 23(2):295–308
13. "Sensichips." [Online]. Available: www.sensichips.com
14. Bourelly C, Ferdinandi M, Molinara M, Ferrigno L, Simmarano R (2020) Chemicals detection in water by sensiplus platform: current state and ongoing progress. In: Di Francia G et al (eds) Sensors and microsystems. Springer International Publishing, Cham, pp 381–386
15. Nopens I, Capalozza C, Vanrolleghem PA (2001) Stability analysis of a synthetic municipal wastewater. University of Gent, Belgium, Department of Applied Mathematics Biometrics and Process Control

# Detect Anomalies in Photovoltaic Systems Using Isolation Forest (Preliminary Results)

S. Ferlito, S. De Vito, and G. Di Francia

**Abstract** Solar energy from Photovoltaic (PV) systems is one of the greatest growing renewable sources of energy. PV systems, although generally quite robust, are subject to failures that can adversely affect energy conversion or even pose safety concerns. Being able to promptly and automatically detect fails/anomalies is essential to improve PV systems reliability while maintaining the expected efficiency. In this paper are investigated five unsupervised Machine Learning (ML) methods for PV plants' faults detection. The tested methods include Isolation Forest (IF), one-Class Support Vector Machines, Local Outlier Factors, Deep Learning Autoencoders, Gaussian Mixtures Models (GMM). Dealing with anomalies in PV means to handle a highly unbalanced dataset, as anomalies are relatively rare events, hopefully. Moreover, it is usually quite challenging to have precise labels for each fault event. For this reason in this paper are analysed only unsupervised ML models, that does not require labelled data. However, being able to quantify models' performance precisely is a quite challenging task as it requires expert support or labels existence, even if they can be imprecise. To accurately represents the model's performance for such a highly unbalanced dataset are reported metrics more suitable for such a task as balanced accuracy, recall, F1, Matthews Correlation Coefficient and Cohen Kappa. For the reason outlined above regarding the available labels, the results summarised in this paper can be considered as preliminary and require a more suitable dataset with precise labels. These preliminary results show that GMM could be highly effective to operate in the PV anomaly detection field; however, some models as IF and Autoencoders, that have proven to be very effective in different but demanding fields, deserve further investigation.

**Keywords** Anomaly detection · Isolation forest · Autoencoders · Gaussian mixture models · Ensemble · Photovoltaic systems

S. Ferlito (✉) · S. De Vito · G. Di Francia
ENEA, DTE-FSD-SAFS, Research Centre, P.le Enrico Fermi 1, 80055 Portici (Naples), Italy
e-mail: sergio.ferlito@enea.it

G. Di Francia and C. Di Natale (eds.), *Sensors and Microsystems*,
Lecture Notes in Electrical Engineering 753,
https://doi.org/10.1007/978-3-030-69551-4_31

# 1    Introduction

Worldwide attention to sustainability has increased the interest into photovoltaic (PV) systems during last years for the unquestioned advantages related to this type of energy source: is almost pollution-free, worldwide available, can be easily integrated onto an existing building. According to IHS Markit [1], globally, new installations of PV systems will reach 142 gigawatts (GW) in 2020, with a rise of 14% over the past year. Cost data reported from International Renewable Energy Agency (IRENA) states [2] a Levelized cost of electricity (LCOE) from utility-scale solar PV of USD 0.068 per kilowatt-hour (kWh) for the year 2019. PV systems, although quite robust and affordable, can incur into many different types of failures that can affect different components that build up such a system. This failure can be incipient, intermittent or lead to permanent failure of a component. A complete analysis of failures affecting a PV system is reported in [3–5]. A list of intermittent faults includes shading, dust, snow accumulation, high humidity. Incipient faults are degradation, corrosion, partial interconnection damage. Among permanent faults, can be enlisted line-to-line, open circuit, bypass diode, arc, junction box. In many cases, PV faults lead, at least, to an energy loss, but can, in some cases, also lead to a risk of fire, for these reasons, monitoring ad automatic fault detection are necessary to timely detect anomalies/faults. Current methods reported in the literature to detect anomalies/faults in the PV system can be roughly divided into three main categories. Visual methods, as IR imaging, electroluminescence imaging, lock-in thermography; power loss analysis, comparison between measured and modelled system outputs; ML-based methods. Many papers employing ML methods have been recently published to detect anomalies and/or trying to isolate/classify them. In [6] is used an approach based on Random Forest ensemble to identify and isolate some faults, including line-line faults, degradation, open circuit, and partial shading using only the real-time operating voltage and string currents of the PV arrays. A common problem in the field subject of interest is that the available dataset is highly unbalanced (few faults event) and often labels (describing the type and/or location of the fault) are often non-available or inaccurate. This problem was overcome in [6] using three datasets: one artificial, built with a Matlab—Simulink based model, where some faults can be ad-hoc created (resulting in this way with accurate labels), one from a small real plant, where some faults are reproduced, and a real production plant with vague labels. This approach, in a similar way, was also considered in [7]. In this case, an unsupervised model, one-class Support Vector Machine (oneSVM) is utilised to detect temporary shading. Authors models a real PV system using a one-diode model then adopt oneSVM method to residuals from the simulation model for fault detection. In [8], an approach that combines Wavelets Multiscales (WM) transforms, and an exponentially-weighted moving average (EWMA), named WM-EWMA is put on test. A single diode model is developed to simulate the functioning of a real 9.54 kWp photovoltaic plant in Algiers. Four types of faults can be detected with the proposed method: open-circuit, short-circuit, temporary shading faults and degradation. Convolutional Neural Network-based model is tested in [9] to detect photovoltaic array fault using only voltage and current of the photovoltaic array as the input features, plus the reference panels used for normalisation. The proposed approach needs a first step where the sequential current and voltage of the photovoltaic array are transformed into a 2-D electrical time series graph to represent the characteristics of sequential data visually. Data concerted in image sequences constitute the input for a Convolutional Neural Network (CNN) deep neural network that performs the photovoltaic array fault diagnosis. A novel deep residual network (ResNet) has been tested in [10]. The proposed method considers PV array's I–V output characteristic curves and ambient condition as input data to build a fault detection and diagnosis (FDD) model able to detect different types and levels of common early faults of PV array such as partial shading, degradation, short and open circuit faults. The raw measured I-V characteristics curves for normal and fault conditions are down-sampled and augmented with irradiance and temperature to build up 2-D data samples. This data is feed to a new ResNet structure which made of residual blocks, convolutional layers, average pooling layers and a linear classifier. In the present paper, authors, investigate ML-based methods to detect anomalies related to power/energy data (not imaging related) acquired from a PV residential plant. The aim is to be able to deal with incomplete or imprecise data trying to only detect anomalies without classifying or isolating the specific type of anomaly/fault. These limitations are due to the specific dataset available (more on this below).

## 2  Dataset Description

The dataset considered in this paper concerns a PV plant located at south Italy with a Nominal Power: 2038.26 (kWp) equipped with a Gefran Inverter of 330 (kW). Data is available from 2018-Jul-19 10:35 to 2019-Aug-31 00:00 (407 days, 109,966 records), data sample every 5 min. Raw fields are datetime, energy (kWh), power (kW), modules' temperature ($°C$), irradiance (W/m$^2$) and a generic report that show for which day a fault with technical support has been recorded. In total have been recorded 14 faults days. Labels indicating a fault (represented as 1) or normal functioning (represented as 0) are retrieved from the available reports by setting to 1 the whole day for which is available a report indicating a generic fault, this is due to the vague/inaccurate reports available. A minimal feature engineering has been applied by retrieving the following features: season (encoded as a categorical feature using ordinal encoding), hour, month and day of the year (all as numeric) to take into account seasonality into the model. With the method outlined above, the proportion of faults/anomalies available in the entire dataset is 1481/109,966 = 0.0013 (1.3%). Kernel density plots for plants DC power for some hours and by season are shown in Fig. 1a, b.

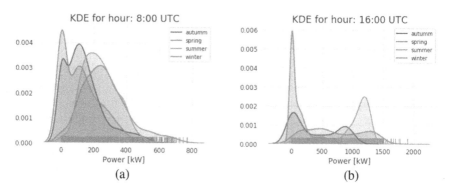

**Fig. 1** **a** and **b** Density plot for plants DC power

## 3 Proposed Fault Detection Models

Five unsupervised ML models have been tested to detect anomalies/faults in PV systems:

1. Isolation Forest (IF)
2. One-Class Support Vector Machine (oneSVM)
3. Local Outlier Factor (LOF)
4. Deep neural network Autoencoders
5. Gaussian Mixture Models (GMM).

Isolation Forest is an ensemble method based on Extremely Randomised Random Forest ensemble model [11, 12]. It is an unsupervised model, computationally fast, able to directly detect anomalies/faults. Moreover, Isolation Forest is quite robust regarding its hyperparameters tuning. IF builds up an ensemble of trees, these trees are built by recursively selecting a random feature and performing random partitioning until an instance is positioned in a terminal node, which means it is isolated. A smaller number of partitions needed to isolate an instance indicates higher chances of an anomaly. The average path length determines anomaly score to the terminal node a given instance has been placed in. In this paper has been evaluated the scikit-learn [13] python implementation of IF, it returns a -1 score for anomalies/faults and 1 for normal data. As this model directly detect anomalies, it can perform worst if the training data does not contain at least a few of them. An oneSVM is an unsupervised learning algorithm that is trained only on the 'normal' data, it learns the boundaries of these points, and it is, therefore, able to classify any points that lie outside the boundary as an anomaly. It does not require labelled data but, as it is required to be trained only on 'normal' data, it performs better if at least is possible to indicate such a type of data. Among the five tested methods, it is computationally the most demanding, and this makes it not ideal for on-line live detection application. LOF is a density-based method for anomaly detection [14], derived from Density-based spatial clustering of applications with noise (DBSCAN); the intuition behind the approach is that the density around an outlier/anomaly will be significantly different from the density around its neighbours. As IF and oneSVM, this method provides an anomaly score, that in scikit-learn implementation assume -1 for faults and 1 for normal data. This method is computationally fast as IF and does not require to be trained on normal data only as oneSVM. Autoencoders are a specialised type of deep neural network whose main goal of training is to provide an output that is the same as the input. They work by compressing the input into a latent-space representation and then reconstructing the output from this representation. The intuition behind autoencoder for anomaly detection is that if trained on 'normal' data they can extract the relevant information for normal functioning and when the reconstruction error is above a specified threshold can detect anomalies/faults. On the contrary of the methods reported so far, this method does not provide an anomaly score, and by carefully select the threshold in the reconstruction error is possible to find an optimal balance between precision and recall. As stated for oneSVM, also Autoencoders need to be trained on 'normal' data. Even if a model based on Autoencoders is an unsupervised one (does not require labelled data), a rough indication for normality is necessary; otherwise, its performance suffers. Many different variations in architecture are available for Autoencoders. In this paper, a simple fully connected deep neural network implemented in $H_2o$ framework (https://h2o.ai) has been employed. Gaussian Mixture Models [15] is an unsupervised clustering method; it is a parametric probability density function represented as a weighted sum of gaussian component densities. Unlike models as K-means, a GMM can give statistical inferences of the underlying distributions; this can be used to calculate the degree of normality or not for data samples (specifying a threshold). In this paper, the scikit-lego (https://sci kit-lego.readthedocs.io/) implementation of GMM has been tested. Like most of the models tested, even this one returns an anomaly score equal to −1 for anomalies and 1 for normal data. As far as the authors know, models as IF, Autoencoder, GMM and LOF have not been yet tested in the field of anomalies in PV systems.

## 4 Methodology and Model's Performance Metrics

The entire dataset has been split in a training and test set using a stratified methodology to guarantee the same proportion of normal ad faults data for both the data partitions. In this way, both the resulting training and test set reflect the same proportion of faults (0.13%) of the whole dataset. All five models under test have been trained on the training set only by performing a k-fold (k = 5) grid-search cross-validation (CV) with F1 metric set as the loss to optimise. Moreover, being the dataset highly unbalanced, for each model, the k-fold grid-search CV described above has been applied to the raw unbalanced data but also to a balanced dataset obtained using one of the four techniques listed below:

1. Synthetic Minority Over-sampling TEchnique (SMOTE)
2. Adaptive synthetic sampling ADASYN
3. Random Oversample

4.  BorderLine SMOTE.

After that, each model, with the optimal parameters obtained by k-fold grid-search CV (trained on the whole training set) have been tested on the unbalanced test set to evaluate their performance on unseen data. The metrics adopted to assess and compare the model's performance need special consideration to evaluate the model's results correctly. Being the dataset unbalanced with a vast amount of normal data is easy to get high accuracy, but this is misleading. Better metrics for this type of scenarios are balanced Accuracy, F1, recall, Mathews correlation coefficient and Cohen's kappa.

$$Balanced\ Accuracy = \frac{\frac{tp}{tp+fn} + \frac{tn}{tn+fp}}{2} = \frac{specificity + sensitivity}{2}$$

This metric takes into account both classes in equal measure and is easily interpretable.

F1 combines precision and recall into one metric by calculating the harmonic mean between those two; it is a particular case of the more general function $F_{beta}$.

$$F_{beta} = (1 + \beta^2) \frac{precision * recall}{\beta^2 * precision}$$

F1 was the metric chosen as function loss optimised in grid-search CV to have a model whose hyperparameters results in a balance between precision and recall. When choosing beta in $F_{beta}$ score, the more one cares about recall over precision; the higher beta should choose. As a result, with F1 score, we care equally about recall and precision while with F2 score, for example, recall is twice as important.

The recall is defined as:

$$recall = TPR = \frac{tp}{tp + fn}$$

It is the go-to metric when interested in anomalies/fault when one really cares about catching all faults even at the cost of false alerts, usually chosen not as only metric but coupled with precision. Matthews Correlation Coefficient (MCC) is a correlation between predicted classes and ground truth; it can be calculated based on values from the confusion matrix:

$$MCC = \frac{tp * tn - fp * fn}{(tp + fp)(tp + fn)(tn + fp)(tn + fn)}$$

MCC is a synthetic metric easy interpretable well fitted to imbalanced problems as is this the case. Cohen Kappa metric tells how much better the model over the random classifier that predicts based on class frequencies.

$$Cohen\ Kappa = \frac{p_0 - p_e}{1 - p_e}$$

**Table 1** Models' anomaly detection accuracy results

| Model | Metric | | | | |
|---|---|---|---|---|---|
| | Acc$_{balanced}$ | F1 | Recall | MCC | Cohen Kappa |
| IF | 0.501 | 0.013 | 0.011 | 0.002 | 0.002 |
| oneSVM | 0.489 | 0.018 | 0.077 | −0.008 | −0.005 |
| LOF | 0.517 | 0.031 | **0.142** | 0.013 | 0.008 |
| Autoencoder | 0.511 | 0.042 | 0.025 | **0.055** | **0.039** |
| GMM | **0.539** | **0.078** | 0.097 | **0.064** | **0.063** |

Metrics (see definitions reported above) to evaluate models' performance

where: $p_o$ is the "observed agreement" and $p_e$ is the "expected agreement". Observed agreement ($p_o$) is how our classifier predictions agree with the ground truth, which means it is just accuracy. The expected agreement ($p_e$) is how the predictions of the random classifier, that samples according to class frequencies, agree with the ground truth, simply put, the accuracy of the random classifier. Values $\leq 0$ mean no agreement, 0.01–0.20 as none to slight, 0.21–0.40 as fair, 0.41–0.60 as moderate, 0.61–0.80 as substantial, and 0.81–1.00 as almost perfect agreement.

# 5 Results and Discussion

Are summarised in Table 1 the results for all five models tested in this paper. As outlined in Sect. 4, all models have been trained both on raw training data and on their balanced version obtained with one of the four balancing methodology reported above. In no case, balancing the training set produced an appreciable improvement in the test set, and the corresponding results are omitted for brevity on the following table. This evidence is probably due to the imprecise labels available for the dataset under investigation and deserve further tests with a more significative and larger dataset. From Table 1 can be derived that GMM, overall, performs better, followed by LOF and Autoencoder. Considering recall, the best performing model is LOF.

# 6 Conclusion

The preliminary results herein reported show that GMM is a promising model/methodology to deal with anomalies in PV with vague or missing labels. Minority class oversampling methods as SMOTE seems to be ineffective. What found is an expected result for models as IF that are able, by nature, to handle highly unbalanced datasets effectively. The same does not hold true for models as oneSVM and Autoencoders. Due to the limitations of the dataset under examine (relatively short with imprecise labels, only DC power and no I-V values for each array in the

PV system), the results reported deserve further investigation. In particular, models like IF and Autoencoders, never tested in PV anomalies task (as far as authors know), have proven to be highly effective in very demanding fields as anti-monetary laundering or telecommunications faults detection and for this reason, have to be more deeply tested with a completer and more accurate dataset.

# References

1. IHS-Markit-Predictions-for-the-PV-industry-2019
2. Renewable Power Generation Costs in 2019. /publications/2020/Jun/Renewable-Power-Costs-in-2019
3. Mellit A, Tina GM, Kalogirou SA (2018) Fault detection and diagnosis methods for photovoltaic systems: a review. https://doi.org/10.1016/j.rser.2018.03.062
4. De Benedetti M, Leonardi F, Messina F, Santoro C, Vasilakos A (2018) Anomaly detection and predictive maintenance for photovoltaic systems. Neurocomputing 310:59–68. https://doi.org/10.1016/j.neucom.2018.05.017
5. Abdulmawjood K, Refaat SS, Morsi WG (2018) Detection and prediction of faults in photovoltaic arrays: a review. In: Proceedings—2018 IEEE 12th international conference on compatibility, power electronics and power engineering, CPE-POWERENG 2018 (2018). https://doi.org/10.1109/CPE.2018.8372609
6. Chen Z, Han F, Wu L, Yu J, Cheng S, Lin P, Chen H (2018) Random forest based intelligent fault diagnosis for PV arrays using array voltage and string currents. Energy Convers Manag 178:250–264. https://doi.org/10.1016/j.enconman.2018.10.040
7. Harrou F, Dairi A, Taghezouit B, Sun Y (2019) An unsupervised monitoring procedure for detecting anomalies in photovoltaic systems using a one-class support vector machine. Sol Energy 179:48–58. https://doi.org/10.1016/j.solener.2018.12.045
8. Harrou F, Taghezouit B, Sun Y (2019) Robust and flexible strategy for fault detection in grid-connected photovoltaic systems. Energy Convers. Manag. 180:1153–1166. https://doi.org/10.1016/j.enconman.2018.11.022
9. Lu X, Lin P, Cheng S, Lin Y, Chen Z, Wu L, Zheng Q (2019) Fault diagnosis for photovoltaic array based on convolutional neural network and electrical time-series graph. Energy Convers Manag 196:950–965. https://doi.org/10.1016/j.enconman.2019.06.062
10. Chen Z, Chen Y, Wu L, Cheng S, Lin P (2019) Deep residual network based fault detection and diagnosis of photovoltaic arrays using current-voltage curves and ambient conditions. Energy Convers Manag 198:111793. https://doi.org/10.1016/j.enconman.2019.111793
11. Liu FT, Ting KM, Zhou ZH (2008) Isolation forest. In: Proceedings—IEEE international conference on data mining. ICDM 413–422. https://doi.org/10.1109/ICDM.2008.17
12. Liu FT, Ting KM, Zhou Z-H Isolation-based anomaly detection
13. Pedregosa Fabianpedregosa F, Michel V, Grisel Oliviergrisel O, Blondel M, Prettenhofer P, Weiss R, Vanderplas J, Cournapeau D, Pedregosa F, Varoquaux G, Gramfort A, Thirion B, Grisel O, Dubourg V, Passos A, Brucher M, Perrot Édouardand M, Duchesnay É, Duchesnay Edouardduchesnay F (2011) Scikit-learn: machine learning in Python
14. Breunig MM, Kriegel H-P, Ng RT, Sander J (2000) LOF: identifying density-based local outliers
15. Li L, Hansman RJ, Palacios R, Welsch R (2016) Anomaly detection via a gaussian mixture Model for flight operation and safety monitoring. Transp Res Part C Emerg Technol 64:45–57. https://doi.org/10.1016/j.trc.2016.01.007

# Sustainable Graphene-Based Mortar and Lightweight Mortar Composites

Giuseppe Cesare Lama, Ferdinando De Luca Bossa, Letizia Verdolotti, Barbara Galzerano, Chiara Santillo, Brigida Alfano, Maria Lucia Miglietta, Tiziana Polichetti, and Marino Lavorgna

**Abstract** "Compact and lightweight" graphene-based mortars were produced and their piezo-resistive behavior was analyzed. Such property is exploitable in the production of functional building materials, which could work as stress–strain sensors. First, graphene-isopropyl alcohol dispersion was synthesized through Liquid Phase Exfoliation and the few layered structure was evidenced via Raman spectroscopy. Afterward, "compact and lightweight" mortars-graphene based composites were produced by using, respectively, cement and cement/diatomite as matrix, the previously mentioned graphene dispersion and a suitable amount of water. The morphological, structural, thermal and electro-mechanical properties of the obtained materials were analyzed. The mix-design here discussed can pave the way for a new kind of eco-friendly, multi-applicative, multi-responsive building material.

**Keywords** Stress/strain sensors · Graphene · Cement-based mortars

## 1 Introduction

Cementitious-Graphene based composites materials have been attracted considerable attention from the scientific and industrial community due to the synergistic effects, mechanical, thermal, durability as well as electrical properties, that the graphene (or its derivatives) is able to induce on the cementitious conglomerates when a percolate and/or segregated path within the matrix is created.

Due to their specific surface area, high aspect ratio, intrinsic physico-chemical properties (e.g., hydrophilicity), graphene [1, 2] and its derivatives [3–6] are able to affect the chemistry of cement hydration [7]. For instance, their intrinsic properties

G. C. Lama · F. De Luca Bossa · L. Verdolotti (✉) · B. Galzerano · C. Santillo · M. Lavorgna
Institute for Polymers, Composites and Biomaterials (IPCB—CNR), P.le E. Fermi 1, Portici (NA), Italy
e-mail: letizia.verdolotti@cnr.it

B. Alfano · M. L. Miglietta · T. Polichetti
Energy Technology Department—Innovative Device Laboratory, ENEA C.R. Portici, P.le E. Fermi 1, Portici (NA), Italy

© The Author(s), under exclusive license to Springer Nature Switzerland AG 2021    239
G. Di Francia and C. Di Natale (eds.), *Sensors and Microsystems*,
Lecture Notes in Electrical Engineering 753,
https://doi.org/10.1007/978-3-030-69551-4_32

promote and regulate the formation of a well-ordered crystalline structure, smaller pores and more uniform pore size distribution leading to significant improvements in mechanical performances, durability (including corrosion resistance due to the reduction of penetration depth of chloride), volume stability (including the autogenous, plastic and drying shrinkage) as well as electrical properties [7, 8].

Together with these aspects, cement-based materials present an important issue for the environment. There are many studies in which researchers, as well as industries, are working for the development of new kind of building material [9–14] In this way "sustainable constructions" can be promoted with the following potential contributions: (a) reduction of cement amount; (b) reduction of construction cycle times due to high early strength; (c) improving thermo-acoustic insulation and environmental properties [15, 16]; (d) stimulating novel "multifunctional" architectural and structural design; (e) providing strong electromagnetic interference shielding property to reduce electromagnetic emission problems on our health.

Usually, the mortar formulation includes cement, sand and water [17]. In this work, we aimed to build up a new kind of mortar, in which the traditional components were replaced by different materials, each one chosen for particular features. In particular, the pozzolana powder (P), which partly substitutes the cement (C), was selected for its natural origin and ease of extraction, as well as its pozzolanic activity [18], in order to let it act as binder. The pozzolana was used leaving its chemistry unaltered, by simply drying the material and sieving the resulting powder, in order to select the powder fraction lower than 100 μm. The milled waste glass (M), coming from the glass industries was used to substitute the sand, it was labelled CPM. In this case, the dimensions of the particles were ranging between 1 mm and 100 μm. This material thanks to its amorphous phase results more reactive with respect to the traditional ones.

Then, in order to increment the bio-sustainability and to reduce the density with the aim to improve the insulating performance of the produced mortar, some samples were prepared by using diatomite powder (D) [19, 20], a nanoporous natural filler silica-based, to partly substitute the sand. This sample was labelled CPM-D.

Other systems, both without and with diatomite, were prepared by using graphene dispersion (0.1 mg/ml), in substitution of part of the' water, whose composition includes water and isopropyl alcohol in ratio 7:1. In order to avoid any re-stacking of the graphene sheets due to a solvent exchange to pure water [21] the dispersion was used as prepared. These samples were labelled CPM-G and CPM-DG, respectively.

## 2  Experimental

### 2.1  Materials

Portland cement 42,5R (C) and pozzolana (P) were provided by EKORU s.r.l. (Volla, Campania, Italy). Milled waste glass was kindly provided by Stazione Sperimentale del Vetro (Venezia, Veneto, Italy). Diatomite was purchased from Sigma Aldrich.

Graphite flakes were obtained from NGS Naturgraphit GmbH Winner Company (Leinburg-Germany). Isopropyl alcohol (IPA) was purchased from Carlo Erba (RS for HPLC- Isocratic grade). Ultrapure water was obtained with a Type 1 Ultrapure Milli-Q system.

### 2.2  Synthesis of Graphene Dispersion

Graphene nanoplatelets were synthesized from natural graphite powder by a Liquid Phase Exfoliation (LPE) method. The process is a sonication-assisted exfoliation of graphite flakes in hydro-alcoholic solution (water/IPA 7:1 v/v). A 1 mg/ml dispersion of graphite flakes was sonicated in a low power bath (around 30 W) for 48 h. Afterwards, unexfoliated graphitic crystals were separated from the dispersion by centrifugation at 500 rpm for 45 min obtaining a black, homogeneous suspension of few-layer graphene at a concentration of 0.1 mg/ml. The concentration was measured by filtering 10 ml of suspension through 0.45 m porosity nylon filters, drying the filters for one day in vacuum at 30 °C and carefully weighing the filtered mass.

### 2.3  Synthesis of Mortars and Composite Mortars Graphene-Based

Several samples were prepared starting form suitable amount of cement, milled glass, pozzolana, diatomite, $H_2O$ and Graphene dispersion [2, 6, 18, 22] (see the details in Table 1).

All samples were produced by first dry-mixing the powders (cement, pozzolana, milled glass and, when required, diatomite). Once the mixture resulted homogeneous, water was added and mixed in order to be uniformly spread. In case of CPM-G and CPM-DG, suitable amount of graphene dispersion was mixed in the system when water was poured on the solid phase. The wet mixture was then, transferred in a 4 × 4 × 16 cm³ mould, and the whole system was put in a closed vessel, partly filled with water, in order to have a 100% of humidity. After 28 days, the cured samples were removed from containers, demoulded, and put in oven at 60 °C for 24 h. Once finished this period, the samples were cut in 2.5 x 2.5 x 2.5 cm³ and characterized.

**Table 1** Formulation of the prepared samples

| Sample | Cement (%) | Pozzolana (%) | Milled Glass (%) | Diatomite (%) | Graphene (%)[a] | H$_2$O/cement ratio |
|--------|-----------|---------------|------------------|---------------|-----------------|---------------------|
| CPM | 12.5 | 12.5 | 75 | – | | 2.2 |
| CPM-G | 12.5 | 12.5 | 75 | – | 0.001 | 2.2 |
| CPM-D | 12.5 | 12.5 | 37.5 | 37.5 | | 4.8 |
| CPM-DG | 12.5 | 12.5 | 37.5 | 37.5 | 0.001 | 4.8 |

[a]Solid percentage with respect to the total amount of the components

## 2.4 Characterization of Graphene

The structural investigations on the graphene were carried out through Raman spectroscopy, with the use of a Renishaw InVia Reflex instrument, equipped with a @ 514 nm laser source. The acquired spectra are the result of 10 accumulations, performed in backscattering configuration on solid films produced by drop casting the starting solution on SiO$_2$-coated silicon substrate; in order to avoid overheating the sample, the laser was used at 5% of its maximum power ($P_{max}$ = 20 mW). Field-Emission Scanning Electron Microscopy (FESEM/EDS LEO 1530-2 microscope, acceleration voltage 3 kV) was used to characterize the graphene nanoplatelets. The morphological, structural, thermal and electro-mechanical properties of the obtained materials were analyzed.

## 2.5 Characterization of Composite Mortars

Piezo-resistive mechanical properties were characterized by using a universal testing machine (model CMT4304 from Shenzhen SANS Testing Machine Co. China, now MTS-USA) equipped with a 30 kN load cell and the cross-head speed was equal to 10 mm/min. During this testing process voltage was applied in DC mode at 20 V, by means of multimeter Keithley source meter 2450 along the direction of compression. Mechanical compressive properties of produced mortars were assessed applying a compression load to specimens having a square section (volume of each sample was 2.5 x 2.5 x 2.5 cm$^3$). The piezo-resistive tests were carried out at room temperature and humidity.

Thermal conductivity, $\lambda$, was evaluated by C-Therm Technologies (TCi) Analyzer, on cylindrical foams (surface section: 6 × 3 cm$^2$).

Morphological analysis were performed with a scanning electron microscope (SEM) Phenom Pure plus ProX (Thermo Fisher Scientific), operated at 15 kV.

# 3 Results and Discussion

## 3.1 Characterization of Graphene Dispersion

Raman spectra, acquired on solid films realized by drop casting deposition, are depicted in Fig. 1.

In the picture, the spectrum of the exfoliated material is compared to that of the starting natural graphite flakes. Both spectra were normalized with respect to the 2D peak and exhibit the typical features of graphitic materials. In more detail, the shape of the 2D band, noticeably different from that relating to graphite, along with the evolution of its two subcomponents $2D_1$ and $2D_2$, indicates the exfoliation of the material down to less than 5 layers [23]. D band, existent only in exfoliated material, is related to the presence of defects. In the specific case, LPE technique is known to generate mainly edge defects [24], therefore the rise of the D band indicates the reduction of the lateral size of the flakes. The SEM images of graphene films (see Fig. 2) show that the platelets have lateral size ranging from hundreds of nm up to few micrometers.

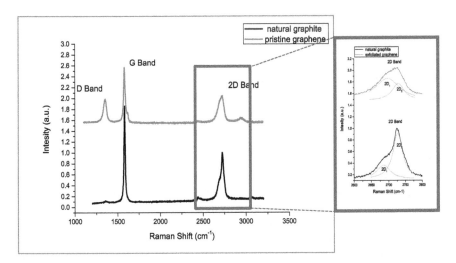

**Fig. 1** Raman spectrum of LPE graphene in comparison with that of natural graphite. In the inset is displayed the deconvolution of the 2D band into the $2D_1$ e $2D_2$ components, indicating a successful exfoliation of graphite down to less than five layers

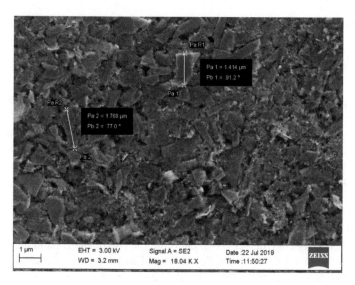

**Fig. 2** SEM micrograph of graphene solid film

## 3.2 Characterization of Mortars and Composite Mortars Graphene-Based

Phenom SEM analyses were performed on the mortar samples and the morphological structure is reported in Fig. 3. As observed the conventional mortar, (CPM, Fig. 3a) highlighted a compact structure with the well-known amorphous phases correlated to the cement and pozzolana phase [18], conversely, in presence of graphene phase (CPM-G, Fig. 3b), the structure is less compact, probably, due to the evaporation of isopropanol. In Fig. 3c, d, e the presence of diatomite [19, 20] in CPM-D, along with the amorphous phase ascribed to the cement and pozzolana was observed. In presence of Graphene dispersion (CPM-DG, Fig. 3f, g, h) the hydrated phases (star-like structure) due to the hydration of cement were observed. In particular, in 3 h, the evidenced hydrated phase results intercalated with a graphene sheet.

In Table 2 the results related to the piezo-resistive mechanical properties along with thermal conductivity properties of the manufactured samples are reported. As observed in Table 2 the addition of graphene filler affected the final properties of both typologies of samples: the compact (CPM) and the lightweight samples (CPM-D). In particular, a reduction in density and mechanical performances (maximum stress and elastic modulus) were observed for the compact mortars, this could be correlated to the presence of the isopropanol of the graphene mixture. In fact, as reported by Kowalczyk et al. [9, 10] the isopropanol has the tendency to reduce the water available to react with the cement powder (with silicate phases, C3S and C2S) and consequently to produce the hydrated phases (Hydrated calcium silicate, CSH) responsible of samples hardening. For instance, the isopropanol acts as a drying agent by removing and selectively replacing water molecules in the interhydrate

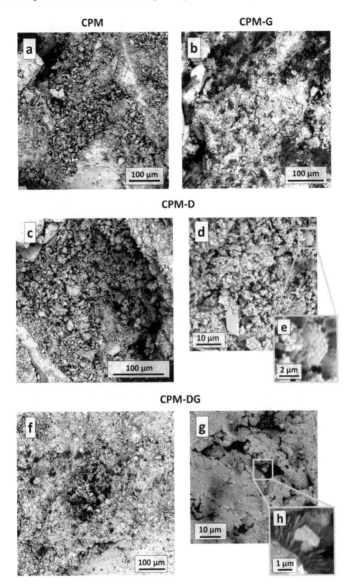

**Fig. 3** Morphological structure of produced mortars

(silicate phases) and capillary network. This leads to a higher quantity of voids, but also, since part of the water was substituted by the isopropanol, to a non-complete hydration of the binder, leading to a less stable microstructure. Furthermore, due to the presence of a higher quantity of voids a reduction in thermal conductivity, $\lambda$, was also detected. Moreover, due to the low amount of graphene filler, the electrical conductivity resulted to be in the same order of magnitude for all samples.

**Table 2** Density, piezo-resistive (Modulus $E_s$, Stress $\sigma_s$, Electrical Conductivity $1/\rho_s$) and thermal conductivity ($\lambda$) analyses results. $E_s$, $\sigma_s$ and $1/\rho_s$ are specific values, evaluated dividing by the density, in order to take into account the porosity of the materials

| Sample | Density ($\rho$, g/cm$^3$) | Modulus $E_s$ (MPa) | Stress $\sigma_s$ (MPa) | Thermal conductivity $\lambda$ (W/K m) | Electrical conductivity $1/\rho_s$ (S/m) |
|---|---|---|---|---|---|
| CPM | $1.404 \pm 0.052$ | $1.394 \pm 0.132$ | $4.688 \pm 0.339$ | $0.294 \pm 0.013$ | $(9.40 \pm 0.61) \times 10^{-6}$ |
| CPM-G | $1.254 \pm 0.047$ | $1.021 \pm 0.089$ | $2.791 \pm 0.245$ | $0.229 \pm 0.011$ | $(6.36 \pm 0.56) \times 10^{-6}$ |
| CPM-D | $0.839 \pm 0.039$ | $0.572 \pm 0.051$ | $1.838 \pm 0.129$ | $0.108 \pm 0.006$ | $(1.90 \pm 0.08) \times 10^{-6}$ |
| CPM-DG | $0.830 \pm 0.040$ | $0.942 \pm 0.073$ | $2.105 \pm 0.189$ | $0.114 \pm 0.009$ | $(0.96 \pm 0.02) \times 10^{-6}$ |

For the lightweight samples (CPM-D), as already described [19, 20], diatomite was chosen for its high intrinsic nanoporosity able to improve the thermal insulation performance of the samples which contain it. This feature has conferred lower density for both diatomite-based systems (CPM-D and CPM-DG) with a consequently halving of $\lambda$ with respect to compact ones. The porous shape of the diatomite microparticles helped to overcome the detrimental effect of the isopropanol on the mechanical properties. They worked as accumulation sites, where the isopropanol was isolated, leaving the water hydrating the binder as also observed in the morphological investigation. In this way, an improvement in mechanical properties (in terms of maximum stress and elastic modulus) was recorded (see Table 2), being the density almost unaltered. However, not significant effects were observed on thermal and electrical conductivity.

# 4 Conclusion

Compact and lightweight composite mortars are produced starting from cement and cement/diatomite matrix respectively and graphene dispersion as conductive nanofiller, together with other eco-sustainable materials.

Despite the low content of graphene, mechanical properties of the lightweight mortars resulted to be improved, where the electrical conductivity was not really affected.

In this view, to produce a piezoresistivity-based strain sensors further studies are required for the optimization of the optimal amount of Graphene conductive nano-particles to improve both the electrical conductivity as well as the mechanical performances on compact mortars and lightweight mortars, trying to reduce the negative influence of the isopropanol on the final system.

**Acknowledgements** Fabio Docimo (CNR-IPCB), Mariarosaria Marcedula (CNR-IPCB) and Alessandra Aldi (CNR-IPCB) are kindly acknowledged for their technical support. Gabriella Rametta (ENEA) is kindly acknowledged for her technical support in SEM imaging of mortar samples. This work has been carried out within the research project "PROSIT - PROgettare in SostenibillTà", funded by POR FESR Campania 2014-2020.

# References

1. Novoselov KS, Geim AK, Morozov SV, Jiang D, Zhang Y, Dubonos SV, Grigorieva IV, Firsov AA (2004) Electric field effect in atomically thin carbon films. Science 306:666–669
2. Castaldo R, Lama GC, Aprea P, Gentile G, Ambrogi V, Lavorgna M, Cerruti P (2019) Humidity-driven mechanical and electrical response of graphene/cloisite hybrid films. Adv Funct Mater 29:1807744
3. Castaldo R, Lama GC, Aprea P, Gentile G, Lavorgna M, Ambrogi V, Cerruti P (2018) Effect of the oxidation degree on self-assembly, adsorption and barrier properties of nano-graphene. Microporous Mesoporous Mater 260:102–115
4. Marotta A, Lama C G, Ambrogi V, Cerruti P, Giamberini M, Gentile G (2018) Shape memory behavior of liquid-crystalline elastomer/graphene oxide nanocomposites. Compos Sci Technol 159:251–258
5. Kuila T, Bose S, Mishra AK, Khanra P, Kim NH, Lee JH (2012) Chemical Functionalization of Graphene and its applications. Prog Mater Sci 57:1061−1105
6. Mugnano M, Lama GC, Castaldo R, Marchesano V, Merola F, del Giudice D, Calabuig A, Gentile G, Ambrogi V, Cerruti P, Memmolo P, Pagliarulo V, Ferraro P, Grilli S (2019) Cellular uptake of mildly oxidized nanographene for drug-delivery applications. ACS Appl Nano Mater 3:428–439
7. Xu J, Zeng W, Chen R, Jin B, Li B, Pan Z (2018) A holistic review of cement composites reinforced with graphene oxide. Constr Building Mat 171:291
8. Dimov D, Amit I, Gorrie O, Barnes MD, Townsend NJ, Neves AIS, Withers F, Russo S, Craciun MF (2018) Ultrahigh performance nanoengineered graphene–concrete composites for multifunctional applications. Adv Funct Mater 28:1705183
9. Hossain MM, Karim R, Hasan M, Hossain MK, Zain MFM (2016) Durability of mortar and concrete made up of pozzolans as a partial replacement of cement: a review. Constr Build Mat 116:128
10. Toghroli M, Shariati F, Sajedi Z, Ibrahim S, Koting ET, Mohamad M (2018) A review on pavement porous concrete using recycled waste materials. Khorami Smart Struct Syst 22:433
11. Letelier V, Tarela E, Muñoz P, Moriconi G (2016) Assessment of the mechanical properties of a concrete made by reusing both: brewery spent diatomite and recycled aggregates. Constr Build Mat 114:492
12. Verdolotti L, Di Maio E, Lavorgna M, Iannace S (2012) Hydration-induced reinforcement of rigid polyurethane–cement foams: Mechanical and functional properties. J Mater Sci 47:6948–6957. https://doi.org/10.1007/s10853-012-6642-5
13. Verdolotti L, Di Maio E, Forte G, Lavorgna M, Iannace S (2010) Hydration-induced reinforcement of polyurethane-cement foams: solvent resistance and mechanical properties. J Mat Sci 45:3388–3391. https://doi.org/10.1007/s10853-010-4416-5
14. Piscitelli F, Buonocore GG, Lavorgna M, Verdolotti L, Pricl S, Gentile G, Mascia L (2015) Peculiarities in the structure—properties relationship of epoxy-silica hybrids with highly organic siloxane domains. Polymer 63:222–229
15. Verdolotti L, Salerno A, Lamanna R, Nunziata A, Netti P, Iannace S (2012) A novel hybrid PU-alumina flexible foam with superior hydrophilicity and adsorption of carcinogenic compounds from tobacco smoke. Microporous Mesoporous Mater 151:79–87. https://doi.org/10.1016/j.micromeso.2011.11.010

16. de Luca Bossa F, Verdolotti L, Russo V, Campaner P, Minigher A, Lama GC, Boggioni L, Tsser R, Lavorgna M (2020) Upgrading sustainable polyurethane foam based on greener polyols: succinic-based polyol and mannich-based polyol. Materials 13:3170

17. Kwan KH, Fung WWS, Wong HHC (2010) Water film thickness, flowability and rheology of cement–sand mortar. Adv Cem Res 22:3–14

18. Verdolotti L, Lirer S, Flora A, Evangelista A, Iannace S, Lavorgna M (2007) Permeation grouting of a fine-grained pyroclastic soil. Proc Inst Civ Eng - Gr Improv. https://doi.org/10.1680/grim.2006.10.4.135

19. Yuan P, Wu DQ, He HP, Lin ZY (2004) The hydroxyl species and acid sites on diatomite surface: a combined IR and Raman study. Appl Surf Sci. https://doi.org/10.1016/j.apsusc.2003.10.031

20. Galzerano B, Capasso I, Verdolotti L, Lavorgna M, Vollaro P, Caputo D, Iannace S,. Liguori B (2018) Design of sustainable porous materials based on 3D-structured silica exoskeletons, diatomite: chemico-physical and functional properties. Mater Des 145:196–204

21. Fedi F, Miglietta ML, Polichetti T, Ricciardella F, Massera E, Ninno D, Di Francia G (2015) A study on the physicochemical properties of hydroalcoholic solutions to improve the direct exfoliation of natural graphite down to few-layers graphene. Mater Res Express 2:03501

22. Salzano De Luna M, Wang Y, Zhai T, Verdolotti L, Buonocore GG, Lavorgna M, Xia H (2019) Nanocomposite polymeric materials with 3D graphene-based architectures: from design strategies to tailored properties and potential applications. Prog Polym Sci 89:213–249

23. Ferrari CA, Meyer JC, Scardaci V, Casiraghi C, Lazzeri M, Mauri F, Piscanec S, Jiang D, Novoselov KS, Roth S, Geim AK (2006) Raman spectrum of graphene and graphene layers. Phys Rev Lett 97:187401

24. Coleman JN (2013) Liquid exfoliation of defect-free graphene. Acc Chem Res 46:14–22

# Laboratory Characterization for Indoor Scenario of the Micro Gas Sensor SGX SENSORTECH MICS6814

E. Massera, L. Barretta, B. Alfano, T. Polichetti, M. L. Miglietta, and P. Delli Veneri

**Abstract** In this paper we present a study on the commercial gas sensor SGX SENSORTECH MICS6814, composed by three microsized MOX, sensors in a controlled environment. Using an innovative and non-conventional approach to the sensor testing, we found a way for the usability of this sensor in an indoor air quality application scenario. As a result we provide a table useful to implement a simple algorithm for the generation of an air quality index that can be displayed with a tachometer or coloured LED.

**Keywords** Microsensor · Gas sensor · IOT · Air quality

## 1  Introduction

Nowadays, people are more and more interested in air quality and demand objects able to monitor and improve air quality. The indoor air purifiers are a class of product that is highly attractive for the market and the future market trends show a sensible growth (>10%) also under conservative expectation [1]. Air purifiers as air conditioning systems are always part of the "House Automation" that means Internet of Things (IOT) objects. Gas sensor equipment on such systems has become a crucial aspect that can determine the market success for these products. For this scenario gas sensors are used to regulate the air purifier system and to provide a quality index of the surrounding air. Industry demands gas sensors to be cheap, microsized, easy to interface to a microcontroller. Each industry dedicated to the gas sensors production has in its catalogue an a-specific microsized sensor, usually based on metal oxides (MOS), for indoor air quality estimation. Air pollutants like ammonia, volatile organic compounds (VOC), ozone, carbon dioxide and carbon monoxide are

E. Massera (✉) · B. Alfano · T. Polichetti · M. L. Miglietta · P. Delli Veneri
ENEA, CR-Portici, P.le E. Fermi 1, 80055 Napoli, Italy
e-mail: ettore.massera@enea.it

L. Barretta
Physics Department "E. Pancini", University of Naples "Federico II", Via Cinthia 21, 80126 Naples, Italy

© The Author(s), under exclusive license to Springer Nature Switzerland AG 2021
G. Di Francia and C. Di Natale (eds.), *Sensors and Microsystems*,
Lecture Notes in Electrical Engineering 753,
https://doi.org/10.1007/978-3-030-69551-4_33

present in indoor air and their concentration levels determine the air quality and the room comfort [2]. The commercial sensor SGX SENSORTECH MICS6814 is one of the most representative of the state of the art of cheap and microsized gas sensors for the indoor air quality indication in an IOT device. As many other devices used in real-world scenarios, environmental gas interferents can affect the stability of the output value of this typology of sensor, causing unreliability and low correlation to the gases concentration [3]. In this work we found that a differential approach is necessary to extract the right information on the air quality surrounding the sensor.

Based on our laboratory tests, we recognize gas sensor features useful for a reliable air quality indication.

## 2 Experimental

### 2.1 he Gas Sensor SGX SENSORTECH MICS6814

The mics-6814 is a compact analog MOS sensor with three fully independent sensing elements on one package. The silicon gas sensor structure consists of an accurately micro machined diaphragm with an embedded heating resistor and the sensing layer on top. The three sensor chips have independent heaters and sensitive layers. Each of these sensors have a particular affinity toward chemicals: one sensor chip detects reducing gases like carbon monoxide or VOCs (sensor called S1), the other sensor is skilled on Ammonia (S2) and the third one detects oxidizing gases like nitrogen dioxide or ozone (S3). The package is compatible with the SMD assembly process (Fig. 1).

The mics-6814 is soldered on a prototypal sensor board produced by the research and development team of MAY S.r.l (Via Francesco Ierace, 5–80,129 Napoli). The Sensor Board is connected with a PC with a USB serial cable and stores LOG files of the sensor digit output (12 bit resolution).

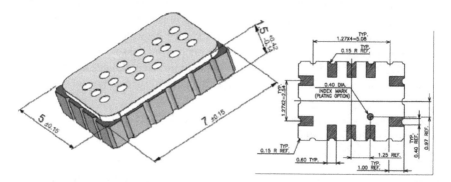

**Fig. 1** Mics-6814 package

## 2.2   Characterization

The Sensor Board was installed in a 15 L box chamber in the Gas Sensors Characterization System (GSCS) developed by Kenosistec Srl. The GSCS ensures a constant temperature and humidity inside the box. Synthetic air in the box chamber is at room pressure and recharged with a constant flow of 1 L/min. Humidity is controlled by mixing dry synthetic air with a humid line in which synthetic air is bubbled in pure water. Air flow, analyte injection and humidity mixing are regulated by high precision Mass Flow controllers (MFC) from MKS. MFC, valves, temperature and humidity, can be set and recorded with a custom software able to run automated tests on sensors. Sensors data LOG were visualized and elaborated with Scripts in R language.

At 21 °C and 50% RH, the gas sensor board was exposed in sequence to:

- Ozone 350 ppb;
- Ammonia 50 ppm;
- Ethanol 50 ppm;
- Carbon Monoxide 50 ppm.

Chemicals and concentrations were chosen to fit a real working environment for the gas sensor.

## 2.3   Test Protocol

For each gas analyte illustrated above, the GSCS run an automatic test protocol composed by three time-step:

*BASELINE* Step—30 min in 1 Lt/min synthetic air.

*ADSORPTION* Step—75 min in 1 Lt/min synthetic air and injection of analyte at the maximum concentration;

*DESORPTION* Step—75 min in 1 Lt/min synthetic air.

Considering the 15 Lt volume of the box chamber in which the sensor board was installed and the 1 Lt/min injection rate of gas analyte at the max concentration, it is possible to precisely estimate (within 5%) the time function of the gas concentration inside the chamber during the test protocol. A previous calibration of the test chamber validated the gas concentration estimation during the test protocol ensuring the appropriate precision and accuracy.

Before each test it was necessary to trim the coupling resistance of the sensors to avoid ADC saturation or too low analog signal. This ruled out the comparison between absolute values coming from different tests.

# 3   Results and Discussion

This work focuses on the description of the scientific methodology applied for the extraction of features from the outputs of unstable sensors.

In Fig. 2. is shown, as an example, the time evolution of the sensor output expressed in digit levels (12 bit, black curve and left axis) in a test with injection of 50 ppm Ethanol (red line and right axis).

The example reported in Fig. 2 is fully representative of the sensor outputs; in particular, a marked drift of the signal in synthetic air was observed for each sensor on the sensor board. After the sensing events, the desorption phase was again affected by the drift. To directly compare different sensors outputs on the same signal scale we subtracted a reference point to each sensor signal. The reference point was set as the time the gas injection started; in this way each sensor signal was referred to its value right before the gas injection. In Fig. 3 is reported the time evolution of the relative response towards 50 ppm of ethanol for the three sensors.

As can be observed in Fig. 3, the drift of the ammonia sensor (S2) seems not affected by the ethanol presence. Actually, this lack of sensitivity was observed even towards ammonia whereas during the test with ozone a change in the rate of the

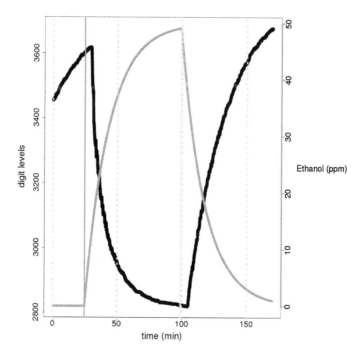

**Fig. 2**  Time evolution of S1 sensor output (digit value) during a 50 ppm ethanol test (black curve). Vertical red line indicates the gas injection point while the vertical green line refers to the injection stop. The red graph refers to the ethanol concentration estimation according to the time function of the analyte concentration in the box chamber (right axis) (Color figure online)

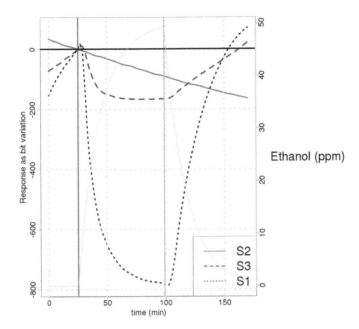

**Fig. 3** Time evolution of the output of the three sensors onboard the mics-6814 during a 50 ppm ethanol test referred to their reference points. The grey graph refers to the ethanol concentration estimation (right axis)

drift was observed (data not shown). This likely indicates a sensor damage occurred during the sensor soldering on the PCB but its responses will be still considered in the following. The other two sensors showed both a negative variation of their outputs but with different intensity. The grey graph refers to the ethanol concentration estimation (right axis).

The instability of the sensor output prevented measuring the concentration level of any of the tested analytes. This very feature is what led us to determine a way to manage this type of output. To this aim, we took into account the rates of variation of the sensors output (VR, bit/min) and plotted these values versus the gas concentration change rate (ppm/min) (Fig. 4).

Figure 4a highlights the different dynamic behaviours of the three sensors towards ethanol. Sensor S1 shows the highest variation rate ($VR_{S1}$) and the ability to detect an increment or decrement under 1 ppm/min of ethanol. Analogous results were obtained towards ammonia, carbon monoxide and ozone and are summarized in Table 1. Another useful feature is the $VR_{S1}/VR_{S3}$ ratio of the variation rate with respect to the contaminant increase as shown in Fig. 4b. Soon after the gas injection a sudden perturbation of the sensor outputs is observed, during which the signal to noise ratio can be too low for one or both the sensors, resulting in a random behaviour of the ratio $VR_{S1}/VR_{S3}$. For the ethanol, this happens between 0 and 1 ppm/min of the adsorption step. Outside this range the ratio usually becomes more stable and,

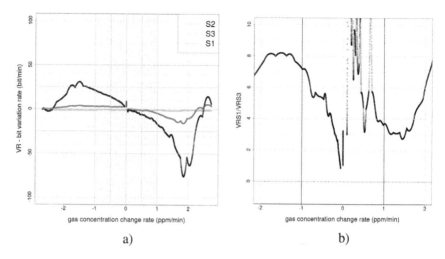

**Fig. 4** **a** Rate of output variation (VR) of the three sensors S1(black), S2 (red) and S3 (blue) related to ppm/min increase or decrease of ethanol during the test. **b** $VR_{S1}/VR_{S3}$ ratio rate variation related to ppm/min increase or decrease of ethanol during the test (Color figure online)

**Table 1** MICS6814—summary of useful features

| Contaminant *max inj* | Ozone 70 ppb | Ethanol 50 ppm | CO 50 ppm | Ammonia 50 ppm |
|---|---|---|---|---|
| Rate of output variation | S3 > S1 > S2 | S1 > S3 > S2 | S1 > S3 > S2 | S1 > S3 > S2 |
| LOD (ppm/min) | 0.003 | 1 | 1.2 | 0.5 |
| S1/S3 in ads ‖ des | > 0.1 ‖ > 0.1 | < 8 ‖ < 8 | < 10 ‖ < 8 | < 5 ‖ < 5 |

for the ethanol test, never exceeded the value of 8. In Table 1 these characteristic values are reported for the different analytes.

In Table 1, the three sensors are listed from the more reactive to the lowest and, for the best performing sensor, we report the limit of detection in ppm/min (LOD) for incremental concentration of the analyte. These values are usually considered on the basis of the signal to noise ratio (S/N) but, in our case, due to marked drift of these sensors, we calculated the LOD of the $VR_{Si}$ in relation to the behaviour of the other sensors of the array. This means that only when the VR ratio of at least one couple of sensors was in the stability zone (Fig. 4b) we could estimate the LOD. This analysis allowed appreciating the characteristic sensitivity of the sensors, as highlighted by the different sensitivity ratio range between sensors S1 and S3 with respect to the different analytes. This can be ascribed to different sensing mechanisms for the two sensors and deserves to be better investigated in the future.

From these indications a simple algorithm can be used to easily communicate an indication on the air quality as illustrated in the following IF AND/OR statements.

Considering $VR_{Si}$ the digit rate variation in bit/min for the sensor i onboard the mics-6814 than it is possible to consider three conditions for the air surrounding the sensor:

- *the air quality is in a stable condition if* $[-15 < VR_{S3} < 25]$ AND $[-15 < VR_{S2} < 25]$ AND $[-15 < VR_{S1} < 25]$;
- *the air quality is worsening*, e.g. air is under contamination by VOCs, if $[-30 < VR_{S3} < -15]$ OR $[-30 < VR_{S2} < -15]$ OR $[-30 < VR_{S1} < -15]$ OR $[25 < VR_{S3} < 50]$ and $\{[-15 < VR_{S2} < 25]$ OR $[-15 < VR_{S1} < 25]\}$;
- *the air quality is becoming bad if* $[VR_{S3} < -30]$ OR $[VR_{S2} < -30]$ OR $[VR_{S1} < -30]$ OR $[VR_{S3} > 50]$ and $\{[-15 < VR_{S2} < 25]$ OR $[-15 < VR_{S1} < 25]\}$.

Moreover if the air is worsening the mics-6814 can discriminate the contamination if the following statement are respected:

- *Ozone* contamination if $[VR_{S3} > 25]$ OR $[VR_{S3} / VR_{S1} > 3]$ OR $[VR_{S3}/VR_{S1} < 0]$;
- *Carbon Monoxide* contamination if $[VR_{S1}/VR_{S3} > 8]$;
- VOCs contamination if $[3 < VR_{S1}/VR_{S3} < 8]$.

These results provide qualitative indications that can be useful in an IOT scenario for the air quality monitoring as feedback for actuators that control air purifiers.


# 4 Conclusion

We have studied the commercial sensor **SGX SENSORTECH MICS6814** installed on a prototypal sensor board in a controlled environment.

Using a non-conventional approach to the sensor testing, we found a way for the usability of this sensor board in an indoor air quality application scenario. In controlled humidity, temperature and time, we exposed the sensor to 70 ppb of ozone and 50 ppm of ethanol, carbon monoxide and ammonia. We measured for each exposure not only the output response of each sensor but also the response speed and the output ratio between the sensors correlated to the rate of gases concentration in the testing chamber. We believe that the ability to detect a change in the concentration of contaminants in an indoor environment can be a useful feature to estimate the air quality and we showed how these features can be used to implement a simple algorithm for an air quality index generation that can be eventually displayed with a tachometer or coloured LED.

These results are worth being further investigated with different sensor boards.

**Acknowledgements** We thank the team of MAY S.r.l for providing the Sensor Board.

# References

1. Wen Y (2017) Air pollution in china and the potential of finnish latest intelligent air purifying technology. Case Lifa Air LA 500. (2017)
2. Wei W, Ramalho O, Derbez M, Ribéron J, Kirchner S, Mandin C, Applicability and relevance of six indoor air quality indexes. Build Environ 15(109):42–49
3. Miletiev R, Iontchev E, Yordanov R (2018) Mobile system for monitoring of air quality and gas pollution. In: International scientific conference on information, communication and energy systems and technologies ICES (2018), pp 28–30

Lightning Source UK Ltd.
Milton Keynes UK
UKHW020643160522
403067UK00006B/496